THE EMBRYOLOGY OF THE HONEY BEE

BY

JAMES ALLEN NELSON, Ph.D.,

Expert, Bee Culture Investigations, Bureau of Entomology
U. S. Department of Agriculture

THE EMBRYOLOGY OF THE HONEY BEE

ISBN 978-1-904846-81-9

Published by
Northern Bee Books,
Scout Bottom Farm,
Mytholmroyd,
West Yorkshire HX7 5JS
Tel: 01422 882751
Fax: 01422 886157
www.GroovyCart.co.uk/beebooks

Printed by Lightning Source UK

PREFACE

The good bee keeper is he who is interested not only in those things which have to do directly with the production of honey but to whom everything pertaining to honey bees has a deep interest. This is shown by the fact that the anatomy of the adult bee has been much studied by practical bee keepers. Aside from the fact that from the egg there emerges in three days a small white larva, no knowledge of the wonderful changes which occur in that small compass is available, while in comparison with the changes which occur, the rather fixed structure of the adult insect seems simple. In this paper is presented to the beekeeping public, as well as to those whose interests are more scientific, the most thorough account of the complex development of the bee egg yet published and to those interested in bees no apology for investigations of this kind is needed. It is of interest to the bee enthusiast, for, while possibly he may not fully appreciate all the details discussed, he will assuredly want to take such facts as his training will permit.

From another standpoint work of this kind is needed. All practical work with bees rests on a foundation of bee activities, to an extent usually not recognized. This is because with few other animals with which man deals is it so imperative that the normal behavior be followed. A study of bee behavior can progress no farther than our knowledge of the structure of the bee has gone and anatomy therefore becomes indirectly a vital thing to the bee keeper. Adult structure cannot be adequately understood without a study of comparative anatomy and especially of development. While the developmental studies are therefore several stages removed from the practices of the apiary, they are nevertheless of importance and the bee keeper will welcome this addition to the foundations of his chosen industry.

E. F. PHILLIPS,

In Charge, Bee Culture Investigations.

TABLE OF CONTENTS

I

HISTORICAL REVIEW

The first recorded observations on the embryology of the honey bee are those of the late Professor Weismann, transmitted by him in a letter to the famous Russian investigator, Prof. Elias Mecznikow (Metschnikoff), who published these in his Embryologishe Studien an Insekten (1866, pp. 489-490). Weismann's statement is as follows: "In the bee a germ layer is formed, which however does not become part of the embryo, but which soon separates from the yolk and becomes an amnion-like envelope. This at first remains in connection with the yolk at the poles, and becomes entirely free only at a later period, when the yolk has become transformed to the true embryo. It is clear that this amnion-like envelope is the embryo, from which there then arises by metagenesis that which we term the bee embryo." These brief and remarkable observations were soon followed by two important papers. The first of these was that of Dr. Otto Bütschli, published in 1870, entitled "Zur Entwicklungsgeschichte der Biene." This paper comprises 45 (8*vo*) pages of text and four double plates. Bütschli studied only the living eggs. The features of the development visible from the exterior were quite accurately described and figured, but the observations of this noted zoölogist were naturally limited by his method, and while he succeeded in discerning correctly the origin and development of certain internal parts, such as the tracheal system, a systematic and detailed account of the origin of the germ layers and of the details of the organogeny was in the nature of the case impossible.

In the next year (1871) appeared an extended memoir by the Russian embryologist Kowalevski, entitled "Embryologische Studien über Würmern und Arthropoden." It comprises seventy pages of text (4*to*) illustrated by twelve plates. This paper is a veritable landmark in the domain of insect embryology. Kowalevski made use of a method new at that time, namely that of cutting sections of the tissue, previously fixed and embedded in paraffin. By the application of this method Kowalevski was able greatly to extend the boundaries of the knowledge relating to the development of insects, and his memoir may be said to inaugurate the period of modern research in this field. In the part relating to the arthropods are included nine pages devoted to the development of the honey bee, illustrated by thirty figures covering plate XI and a half of plate XII. Although that part of the section devoted to the honey bee, which records observations on the living egg, covers much the same ground as that covered by Bütschli in the previous year, nevertheless by the study of sections Kowalevski was able to add many important facts regarding the formation of the blastoderm, the origin and fate of the germ layers, and the development of the nervous system. The figures are excellent, and although small, are sharp and clear, while the text is written in a condensed but lucid style.

Owing probably to the excellence of the observations just described, over ten years elapsed before the egg of the honey bee was again made the subject of scientific investigation. In 1884 a paper by the Italian zoölogist Battista Grassi appeared in a relatively obscure journal—Atti dell' Academie Gioenia di scienzi naturali in Catania—, and according to Carrière (1897) remained for a time almost unnoticed. This paper covers seventy-seven pages of text (4to), and is illustrated by 252 figures covering ten plates. This is the fullest, and in fact the only complete account of the embryology of the honey bee ever published. Like his predecessors Grassi studied the living bee, but he also made free use of microtome sections. The text consists of an introduction including a brief review of literature, twelve numbered chapters, each descriptive of the development of a tissue or system of organs, and in addition a chapter discussing the significance of the facts recorded. This most excellent paper is scarcely open to criticism, when judged by the standard of contemporaneous papers, and allowing for the relative crudity of the histological technique of that time. Indeed the correctness of most of the facts recorded are beyond question. Nevertheless, judged by modern standards, the text seems lacking in completeness, while most of the figures appear crude and diagrammatic, many of them losing much of their value by their small size. In addition the correctness of some of Grassi's statements are open to question.

In this connection it will be necessary merely to mention Blochmann's paper (1889), since it relates only to the maturation of the egg. This may also be said of Petrunkewitsch's paper (1901). A second paper by this investigator (1903), "Das Schicksal der Richtungskörper im Drohnenei" however contains some data and figures relating to the early development of the egg of the honey bee. These will be mentioned later. The recent paper by Nachtsheim (1913) should also be mentioned here. Although this is also concerned principally with the fertilization the maturation of the egg of the honey bee, nevertheless it contains a number of excellent figures of cleavage cells together with a considerable amount of data regarding their cytological features.

The only other paper of recent date devoted to the embryology of the honey bee is that of Otto Dickel (1903) entitled "Entwicklungsgeschichtliche Studien am Bienenei," comprising forty-six pages (8vo) illustrated by forty-six text figures and two double plates. This paper is extremely limited in its scope, describing only the early development up to but not including the formation of the germ layers. It was submitted as a thesis for the degree of Doctor of Philosophy in the University of Munich, and was produced under the supervision of Prof. Oscar Hertwig of that institution. It was apparently written with one end in view, namely to demonstrate that in the honey bee the mid intestine arises from the yolk cells ("entoderm"), and loses much of its scientific value by its ill-concealed attempt to arrive at a predetermined conclusion.

This closes the list of papers descriptive of the embryology of the honey bee, but in this connection one paper should be noticed which is of the highest value to all students of insect embryology, especially that of the Hymenoptera. This is the beautiful memoir of Carrière and Bürger, "Die Entwicklungsgeschichte der Mauerbiene (*Chalicodoma muraria* Fabr.) im Ei." This consists of 165 pages (4to) accompanied by one single plate and eleven double plates, nine of which are colored. This work covers the entire development of the egg of the mason bee, from the commencement of cleavage to the hatching of the larva, and it stands alone as the most complete account of the embryology of a single insect.

II

ORGANIZATION OF THE EGG

In form the egg of the bee approximates a long cylinder; one end, however, being slightly larger than the other, and both having a smoothly rounded hemispherical contour. The egg is gently curved in its long axis, so that in profile one side is seen to be decidedly convex, the other slightly concave (Fig. I). By reflected light the egg appears pearly white, by transmitted light it is seen to be translucent, with a saffron tinge.

In length the eggs vary from about 1.53 mm. to 1.63 mm. (0.059-0.063 inch). The larger end at its broadest point has.an average diameter of about 0.317 mm. (0.0122 inch), or approximately one-fifth of the total length.

Both the differing size of the two ends and the marked bilateral symmetry of the egg, expressed by the curvature of its long axis are common among the eggs of insects in general and always bear a direct and constant relation to the position of the future embryo. In the egg of the honey bee, with reference to the parts of the embryo, the larger of the two ends is cephalic (anterior), the opposite caudal (posterior); the convex side, ventral, the concave, dorsal. The position of the embryo is therefore predetermined. Hence the terms *cephalic pole, caudal pole, dorsal* and *ventral* sides may be applied directly to the egg itself. A similar form and similar relations to the embryo are possessed also by the ova of other Hymenoptera, for example *Formica* (Ganin, 1869), *Chalicodoma* (Carrière and Bürger, 1897), and *Polistes* (Marshall and Dernhehl, 1905). In many other insects it is known, moreover, that the egg before deposition lies in the ovary of the mother in such a position that the parts of the future embryo are directed or oriented coincidently with those of the mother. That is, the cephalic pole of the egg is turned toward the head of the mother. This is known as the law of orientation of Hallez, named from its discoverer (1886).

4

Whether this relation obtains in the bee has so far apparently not been demonstrated. It is at least certain, however, that the egg is generally deposited with the cephalic pole directed outward, or toward the mouth of the cell. If, as is in all probability true, this is the position in which it comes from the ovipositor of the queen, then the future caudal pole of the embryo would correspond with the caudal end of the queen, and, since in the ovary of the queen the eggs are disposed parallel to her long axis, the law of Hallez doubtless applies to the honey bee.

As is known to every bee keeper, the eggs are commonly placed, one in each cell, at or near the center of its floor. The egg is attached by its smaller or caudal end by means of a minute quantity of an adhesive substance secreted by the queen, and it is thus enabled to stand at right angles to the bottom of the cell. While this is the usual manner in which the eggs are deposited, many deviations from it are frequently observed. These take the form of variations in number and in position. Frequently two or more eggs may be laid in a single cell; Grassi records finding as many as six; but this number is frequently exceeded, bee keepers sometimes finding as many as two dozen. Such eggs may or may not be in the same stage of development; they may be laid separately or adhering to one another. In examples of the case last mentioned it is of interest to note that the adhesive area of the egg is not confined to the caudal pole but extends well up toward the middle of the egg. Variations in position are very common; for example, the egg may be placed either on its side, or on its posterior end; against the bottom or the walls of the cell. As far as known such variations as these do not necessarily imply abnormality either in the queen or in the egg, but it is of course impossible for two or more larvae to long continue their development together in a single cell. Probably in most cases, the workers see to it that superfluous eggs are soon removed, although two larvae are occasionally found in a single cell. There is a general belief among bee keepers that such deviations in the position or the number of eggs laid in a cell indicate abnormality in the queen, since they most frequently occur when the queen is unfertilised or senile, or in the case of workers becoming fertile in the absence of a queen and of worker brood. It is of interest to note that, according to Marshall and Dernehl

(1905), the egg of *Polistes* is usually attached to that wall of the cell nearest the center of the nest and about two-thirds of the depth of the cell from its mouth.

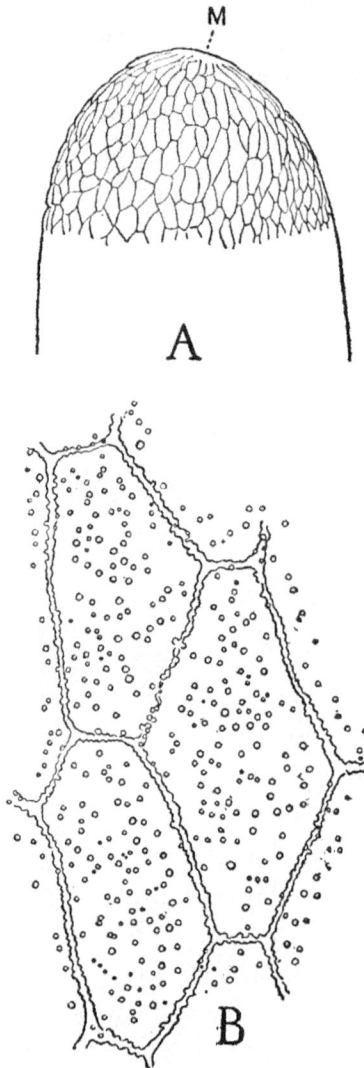

Fig. 1.   A, Surface view of anterior end of egg, showing the reticulated chorion, and the micropylar area (*M*) ; x 100.   B, Surface view of a small portion of the chorion showing reticulum and tubercules, x 713.

Two membranes cover the external surface of the egg.   The outer of these, the chorion, (Fig. 1, A and B), is a very thin and transparent but tough and dense sheet about half a micron in thickness, completely surrounding the egg, and protecting it from direct contact with the atmosphere.   The composition of this membrane has been studied by Tichomiroff (1885) in the silk-worm, more recently by Lecaillon (1897), and by Lecaillon and Henneguy conjointly (1903) in the eggs of the representatives of several orders of insects. These investigators have all found that the chorion is not composed of chitin, as is the integument of the imago, but of a peculiar substance differing in composition from both chitin and horn, which Tichomiroff has termed *chorionin*. The tests employed by these investigators have not been applied to the chorion of the bee's egg, but it is altogether probable that it also is composed of chorionin.

Like other insect eggs the chorion of the bee's egg is not smooth but sculptured. This sculpturing takes the form of delicate ridges forming a meshwork over nearly the entire egg (Fig. 1A). The meshes have the forms principally of pentagons and hexagons elongated in the direction of the long axis of the egg. Examination under high power shows that the ridges forming the polygons (Fig. 1B) are made up of minute circular papillae fused together at their bases, and that similar papillae are scattered about over the area within the polygons. Patterns of this general character are commonly found among insect eggs, the polygons representing merely the imprint of the ovarian follicle cells which secreted the chorion. At the posterior end of the egg the ridges fade out and disappear. At the extreme anterior end the ridges converge toward a small area over which the chorion is conspicuously thickened. This area in general appearance and position corresponds to the micropylar area in other insect eggs and will accordingly be termed such (Fig. 2). It consists of a plate-like thickening of the chorion, approximately circular in outline, whose margins are continuous with the ends of the ridges mentioned above. This thickening exhibits a number of what appear to be perforations or fenestra; these are very irregular in outline and at the edges of the micropylar area merge with the spaces between the ridges.

In other insect eggs the micropyle is a perforation or system of perforations of the chorion permitting the entrance of spermatozoa into the substance of the egg. Frequently the micropyle is very complex in structure, and sometimes closed by a gelatinous plug. It is very frequently, although not always, situated at the cephalic pole of the egg. The writer has not been able to demon-

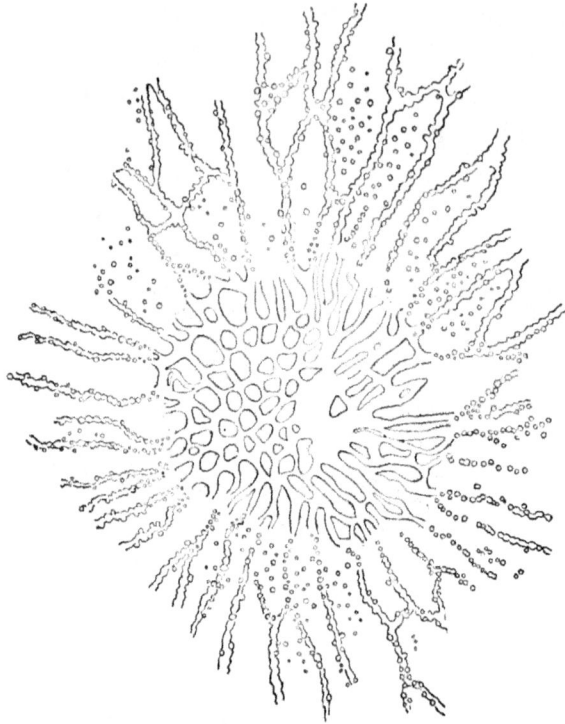

FIG. 2. Micropylar area, as seen in face view, x 713.

strate actual perforations in the micropylar area of the bee's egg.
All of the sections of the eggs have been unsatisfactory in this
regard.   This is in part due to the density and toughness of the
chorion, these qualities being intensified by the process of dehy-
dration preliminary to embedding in paraffin, so that clear and
sharp transverse sections of this portion of the chorion are rare.
The indirect evidence that this is the micropylar region is however very strong.   In the first place, as has already been said,
the micropyle is in very many insect eggs situated at the anterior pole.   In the second place no other portion of the chorion
displays any differentiation which could be interpreted as a
micropylar apparatus; and lastly Blochmann (1889), Petrunkewitsch (1901) and Nachtsheim (1913) have found that the spermatozoa do actually enter the egg at the anterior pole.   In spite
of the lack of direct evidence of the actual perforation of the
chorion at the anterior pole it is fairly safe to assume that this is
in fact the micropylar area.

The vitelline membrane, the second of the two membranes enclosing the insect egg, is in the bee somewhat thinner than the

chorion and apparently structureless. It is sometimes found adhering to the chorion, sometimes to the egg.

The contents of the egg of insects, as well as of all other animals, is made up, broadly speaking, of two portions, namely: *protoplasm*—the so-called formative yolk,—the material basis of all life and the part of the ovum immediately concerned in development; and *deutoplasm,* a store of food material destined to be consumed by the developing embryo. In the vast majority of insect eggs the deutoplasm greatly exceeds the protoplasm in amount.

If an egg of the bee, freshly taken from the comb during the early stages of development, is placed in normal salt solution and crushed slightly so as to rupture the chorion and allow some of the contents of the egg to flow out, examination under the microscope (Fig. 3) will show that these consist of the following

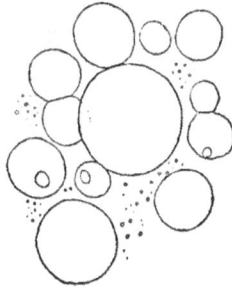

FIG. 3. Yolk from a living egg, showing vitelline spheres, vitelline bodies, and the minute refringent bodies supposed to represent Blochmann's corpuscles, x 600.

elements: (1) Large transparent spheres of a colorless fluid, 10-25 microns in diameter, the *vitelline spheres.* (2) Small refringent bodies, apparently solid, lying within these spheres, the *vitelline bodies.* Their size is variable, but is about one-sixth the diameter of the spheres; their form is also variable, rounded to long ovate, best expressed by comparing them to the water worn pebbles of glacial gravel. Their number is approximately the same as that of the spheres. (3) A viscid interstitial substance, the ovarian protoplasm, pale greenish in color, apparently cementing the spheres together. (4) Multitudes of very tiny greenish refringent bodies lying within the interstitial substance. These may possibly correspond to the so-called *Blochmann's corpuscles* found in certain other insects. They are not distributed uniformly, but are gathered into small groups. These bodies, in contact with the salt solution often display a lively dancing (Brownian?) movement, recalling that of some of the flagellate infusoria. Examination of a similar uninjured egg shows it to be densely packed with the transparent vitelline spheres (1), while surrounding them everywhere and filling the interstices between them is the interstitial substance (3), which also forms a cortical layer around the periphery of the egg. The smaller clear bodies (2) are uniformly distributed throughout the egg, while the tiny greenish Blochmann's corpuscles (4) although present within the interstitial substance throughout the egg, are especially abundant in the cortical layer. Since of these four components of the egg, one, namely the viscid interstitial substance, is protoplasm; the remainder, therefore, with the possible exception of the Blochmann's corpuscles, constitutes the deutoplasm, and this makes up by far the greater volume of the substance of the ovum. In its physical make-up the egg contents closely approximate that of an emulsion.

Turning to sections of fixed and stained ova in the earlier stages of development (Figs. 4A and B, Fig. 5) the egg is seen to be filled with a network of deeply stained protoplasm enclosing irregularly circular spaces. These are largest near the center of the egg and diminish in size near the periphery. Around the latter is a layer of protoplasm continuous with that forming the meshwork. This is the *cortical layer* or "Keimhautblastem" of Weismann and other German investigators (Figs. 4A and 5, *CL*). It has a thickness of about 60 microns near the anterior pole of the egg and diminishes gradually toward the posterior pole to about one-half of this thickness. Near the anterior pole, on the ventral surface, this layer sends out a conical projection into the interior (Figs. I and 5, *PP*). This is the polar protoplasm, the "Richtungsplasma" of Petrunkewitsch. At this stage it contains the polar bodies, or their remains. During the formation of the blastoderm it disappears. Along the central longitudinal axis of the ovum, particularly in its anterior half the strands of the protoplasmic meshwork are much thicker than those nearer the periphery of the egg (*cf.* Figs. 4A and B). Within the spaces of the meshwork are scattered rounded bodies, spherical to long ovoid in form, always more or less densely stained, but showing

an especial affinity for iron haematoxylin. Referring to the
description of the contents of the ovum as seen in the fresh con-

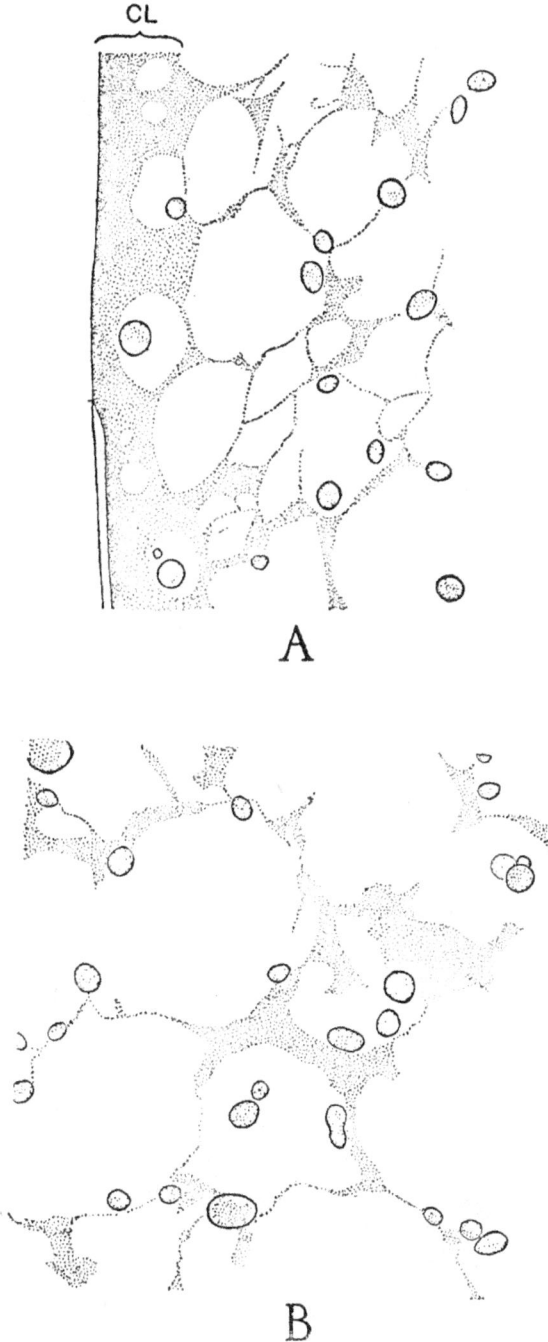

FIG. 4. Protoplasmic network, from sections of unsegmented eggs
fixed in picro-formol. A is taken from the periphery of the egg and shows
the cortical layer (CL); B is taken from near the center of the egg. The
vitelline bodies are conspicuous in both figures, x 1107.

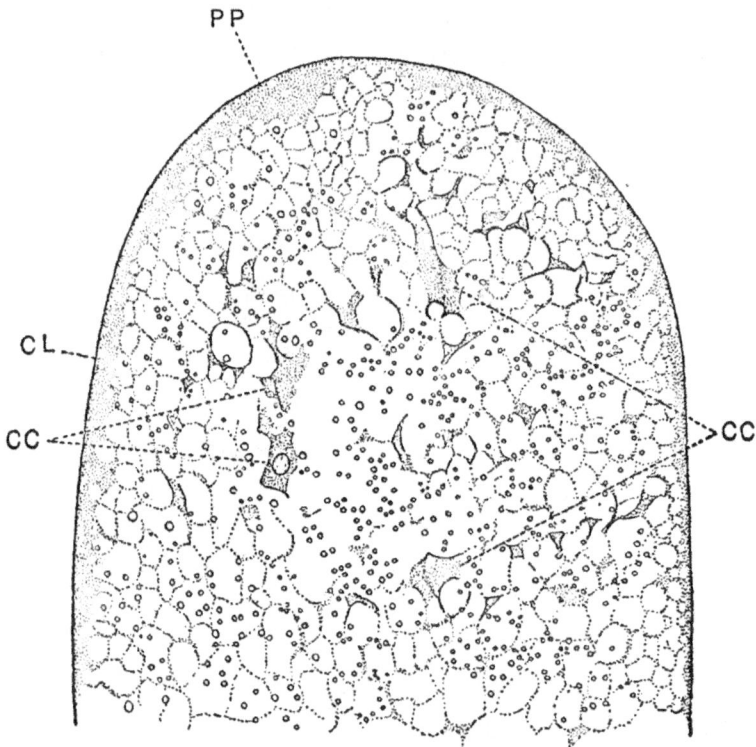

FIG. 5. Anterior half of a sagittal section of an egg, Stage I, showing the polar protoplasm (*PP*), the cortical layer (*CL*) and portions of several cleavage cells (*CC*), x 243.

dition, the deep staining meshwork is readily identified as protoplasm (3), the cavities between the meshes representing the spaces formerly filled by the vitelline spheres. These cavities were of course originally spherical, but have become more or less distorted by the action of the reagents with which the eggs were treated. These spaces are however not invariably empty. In many preparations, more particularly those treated with acetic alcohol or Gilson's fluid, a pale granular precipitate partly or entirely fills these spaces. The small rounded deeply staining bodies are evidently to be identified with the vitelline bodies (2). They are, however, not always present, or rather are present in varying number in different preparations. The natural inference is that they are more or less soluble in the reagents with which the eggs are treated preliminary to sectioning, since they are always visible in fresh material. These bodies are always found in sections of ova fixed with picro-formol—although much more abundant in some preparations than in others—but are sparse and scattered in sections of ova fixed in acetic alcohol, and absent in those fixed in Petrunkewitsch's fluid. So far all the elements seen in the fresh egg have been accounted for except the tiny greenish bodies embedded in the protoplasm (4). These seem to be represented by deeply stained minute granules, which indeed appear to lend to the protoplasmic network its dark appearance.

In addition to the structural elements above mentioned, there are visible in the anterior end of the egg one or more irregular island-like masses composed of the same granular protoplasm as the network, the cleavage cells (Figs. I and 5 CC). Each of these possesses a clear spherical nucleus. At the close of the process of fertilization there is of course but one nucleated cell, whose nucleus (first segmentation nucleus) arose by the union— in the fertilized egg,—of the nuclei of the egg and the sperm (male and female pronuclei).[1] This cell soon gives rise by mitotic division to a group of daughter cells, four of which are more or less plainly visible in the figure. The form of the cleavage cells is highly irregular, or amoeboid, their outlines indented by concavities which represent the imprint of the vitelline spheres. Between the indentations arise processes of irregular shapes which stretch out into the surrounding cytoplasmic network and are continuous with it. These cells will be considered at greater length in the next section.

An extended review of the various accounts of the organization of the eggs of other insects is not possible here; moreover a survey of these accounts, as well as of the statements in the textbooks, suffices to show that they quite uniformly agree in describing the contents of the insect egg as composed of the following morphological elements: (1) A protoplasmic meshwork, which in many insects is extended over the periphery of the egg to form a cortical layer; in others this layer is absent. In the protoplasm are also included the segmentation nucleus or its products. (2) Yolk bodies, generally in the form of balls or spheres, enclosed in the meshes of the protoplasm. (3) Fat globules. (4) Minute rod-like or rounded bodies embedded in the substance of the protoplasm, and present only in certain insect eggs. These bodies were discovered and described by Bloch-

---

[1] For an account of this process in the honey bee, see Blochmann (1889), Petrunkewitsch (1901, 1902), and Nachtsheim (1913).

mann (1884, 1886, 1887, 1892) for the eggs of *Phyllodromia*
(*Blatta*), *Periplaneta*, *Blabera*, *Lasius*, *Pieris*, *Musca*, and
Vespa. In the first four genera named these bodies were rod-
shaped, and strikingly resembled certain bacilli. Moreover, this
resemblance was enhanced by the fact that these bodies multi-
plied by transverse fission. They were found to be especially
abundant in the cortical layer, and also present in the fat body
of the imago. In the three genera last named, these bodies were
rounded or granular in shape and only tentatively identified with
the bacillar form. Wheeler (1889) has also found and described
these bodies in *Blatta,* and gave them the name "Blochmann's
corpuscles." Mercier (1906) has demonstrated that these are
independent organisms, probably bacteria, and capable of culti-
vation in artificial media. Friederichs (1906) has recently described similar bodies, which he terms "Blochmann's corpuscles"
("Blochmannische Körperschen"), embedded in the protoplasmic
meshwork and cortical layer of certain chrysomelid beetles
(*Chrysomela, Rhagonycha*), and Tanquary (1913) finds similar
bodies in the eggs of an ant, *Camponotus*. In the case of the
honey bee the minute bodies to which has been given the term
"Blochmann's corpuscles" are of course comparable only to the
rounded form, and only provisionally identified with these, since
in the honey bee so little is known concerning these bodies that
final identification would be premature.

Turning to the accounts of the hymenopterous egg in particular, there are available the accounts of Henking (1892) for
Lasius, Carrière and Bürger (1897) for the mason bee (*Chalicodoma*) and *Anthophora,* Marshall and Dernhehl (1905) for
Polistes and Tanquary (1913) for *Camponotus;* for the honey
bee Bütschli (1870), and Grassi (1884). Henking's observations contain little of interest here, except that the egg of *Lasius*
possesses a well developed cortical layer. In the case of *Chalicodoma* the account is brief, essentially contained in the statement that the contents of the egg "very fluid, consists of an
emulsion of a considerable amount of deutoplasm (food yolk)
in a small amount of protoplasm." The cortical protoplasmic
layer is wanting, but present in *Anthophora*. The data given by
Marshall and Dernhehl regarding *Polistes* are brief and relate
principally to the cortical layer. This is very similar to that of

the honey bee but is thicker on the ventral than on the dorsal side. In *Camponotus* Tanquary mentions a well developed cortical layer, containing a large number of Blochmann's corpuscles, and having also embedded in it numerous yolk granules. Vacuoles, so called, are present at the center of the egg, more especially near its anterior end. Possibly these correspond to the yolk spheres of the honey bee.

In the descriptions of both Bütschli and Grassi are noted the protoplasmic network, the vitelline spheres, the vitelline bodies and Blochmann's corpuscles. The latter, it is true are not specifically mentioned, but the protoplasm is spoken of in both instances as "granular," an appearance referable to the presence of these bodies. These investigators also observed the cortical layer, describing it also as "granular."

As regards the chemical make-up of the deutoplasmic portion of the insect egg, very little is known. It is commonly described as consisting of yolk and oil globules. This paucity of information is no doubt due to the fact that embryologists usually have not the training requisite to enter the difficult field of microchemistry. It is true that Tichomiroff (1885) made an elaborate chemical analysis of the silkworm egg, but this work has apparently stood alone. Friederichs (1906) subjected the contents of the egg of *Rhagonycha* and *Chrysomela* to a few simple reagents: water, normal salt solution, an aqueous solution of osmic acid and mercuric chloride, and diluted hydrochloric and acetic acids combined. The results convey but little information concerning the real nature of the deutoplasm except to demonstrate the presence of oil droplets. The presence of these appears to be quite general in insect eggs. Grassi assumed that the vitelline spheres in the egg of the honey bee are of an oily nature; a natural supposition, suggested by their appearance. Repeated tests with osmic acid, however, failed to yield the well known blackening reaction characteristic of fat, nor did other elements in the egg show it. The egg of the bee therefore is presumably totally devoid of fat or oil, as such. A few tests with other reagents were applied to the vitelline spheres, but with inconclusive results, except that they appear to indicate that the vitelline spheres in the bee are similar to those in the eggs examined by Friederichs.

III

CLEAVAGE

The group of cleavage cells[2] formed by the division of the segmentation nucleus is at first nearly spherical in outline, and situated in the middle of the anterior end of the egg about equidistant from the cephalic pole and the ventral, dorsal and lateral surfaces. By rapid multiplication and migration of its component cells, this group elongates in a caudad direction until at Stage II (4-6 hours) it extends some two-thirds of the length of the egg toward the caudal pole. The cleavage cells are now arranged in the form of a one-layered meshwork forming a hollow figure whose outline is long conical or pyriform, the slender pointed end directed caudad, the larger rounded end close to the surface at the anterior pole (Fig. II). A cross section through the anterior portion of an egg at this stage is shown in figure 6. The cleavage cells are here seen arranged in a circle whose position is slightly eccentric with respect to the median axis of the egg, being displaced a trifle toward the ventral (lower) side. The cells are still situated at some distance from one another, as well as from the periphery of the egg.

With the change in the form of the group of cleavage cells from spherical to elongate there arises a difference between the cells composing its anterior and posterior portions. The difference consists in this: that the cells in the anterior portion of the cleavage figure are more numerous, more compact in form and

[2] Blochmann (1887a) has called attention to the fact that the complex made up of the cleavage nuclei, together the protoplasm immediately surrounding them, and the protoplasmic meshwork, including the cortical layer, is a syncitium. It is therefore not strictly correct to use the term "cleavage cells" as applied to the cleavage nuclei and the protoplasm immediately surrounding them. The employment of this term may however be justified both on the ground of convenience, and because the cell territory appertaining to each nucleus is always distinguishable, except for a very brief period during the formation of the blastoderm. This position is held by Heider (1889), Heymons (1895), Lecaillon (1897), etc.

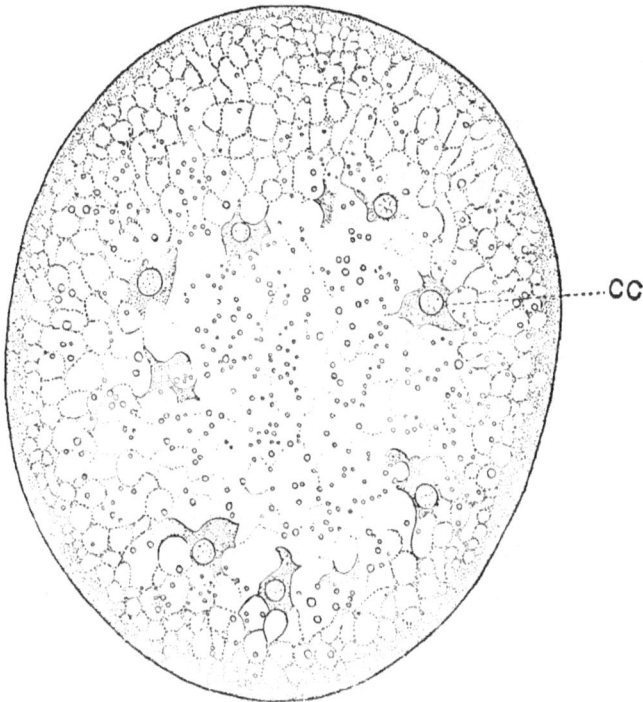

Fig. 6. Tranverse section of anterior end of egg of Stage II, showing cleavage cells (*CC*) x 243.

joined more closely together than those situated further caudad; the latter present by comparison a scattered and somewhat attenuated appearance. This difference is plainly evident in figures II and III (*cf.* also Figs. 7A and 7B). Moreover, although all the cells in the egg are of the same age those in the posterior portion are retarded in development in proportion to their distance from the anterior end, and since they reach their destination in the cortical layer later than those situated anterior to them, a handicap, so to speak is placed on the posterior region of the egg, from which it is slow to recover. This condition has also been noted by Carrière and Bürger for *Chalicodoma,* but is entirely ignored in Dickel's account (1904). Marshall and Dernhehl also make no mention of it in the case of *Polistes,* nor does it appear in *Lasius* (Henking).

As the number of cells continues to increase the pyriform figure grows both in length and diameter until it finally touches the cortical layer of the egg at one or more points. The point where the cleavage cells first actually come into contact with the periphery of the egg is usually situated on the ventral side, one-

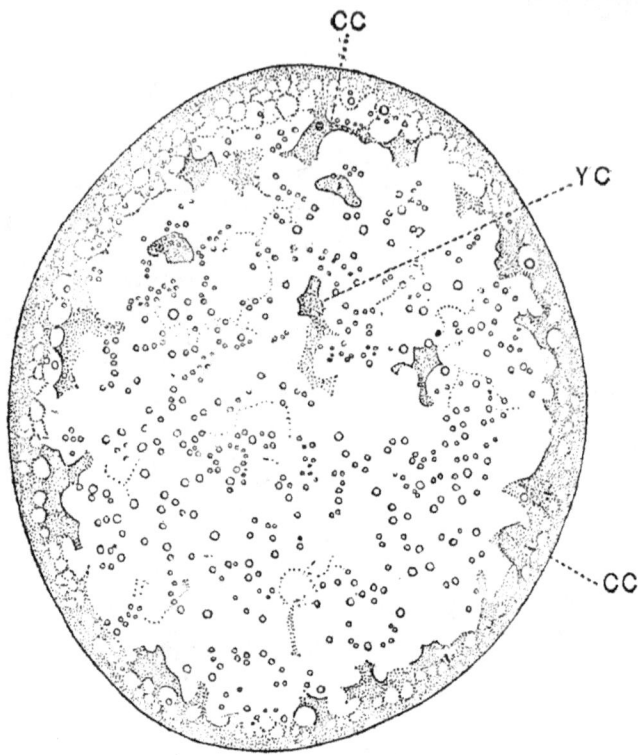

A

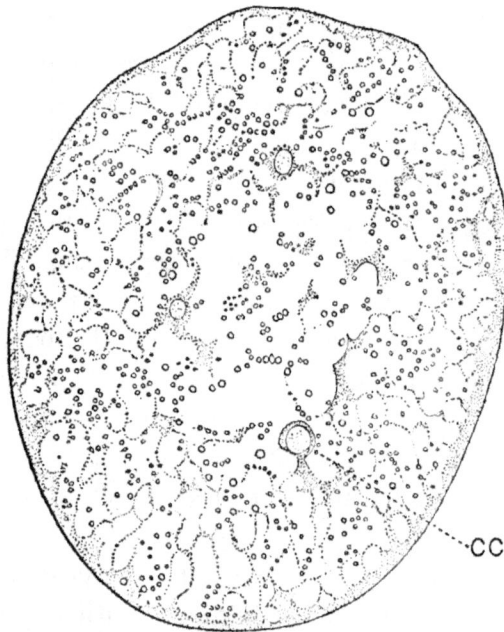

B

Fig. 7. Transverse sections of egg of Stage III. A, from anterior half, showing cleavage cells (*CC*) near the periphery, and yolk cells (*YC*) near the center. B, from posterior end. showing cleavage cells (*CC*), x 243.

fourth to one-third of the length of the egg from the cephalic pole. In some cases the cells appear as though concentrated on this point, but its location is probably subject to considerable individual variation. Soon after the first cells reach the surface the anterior half of the egg becomes rapidly invested with a covering of cells, yet the ventral surface is, as figures III and 7A show, generally attained a little before the dorsal. The attainment of the periphery of the egg by the cleavage cells now progresses slowly caudad, the caudal pole itself being the last to receive a cellular investment.

By far the greater number of cells advance toward the periphery in a single row, eventually attaining the cortical layer and then forming blastoderm, nevertheless, a few linger behind within the yolk and become the yolk cells. Three of these are clearly seen in figure 7A, *YC*. At this period they are not distinguishable from the other cells by any visible characteristics. Their origin has already been correctly described by Dickel (1904) for the honey bee, and conforms to that of the higher insects in general. The further history of these cells will be considered later.

The varied and highly irregular amoeboid contour of the individual cleavage cells is illustrated by figures II, III, 6, 7A and B, as well as by the series represented in figures 8A-F. As the figures show, the outline of the cell is indented by concavities; these concavities represent the spaces occupied by the vitelline spheres. Between these concavities irregular slender processes extend out from the cell; these either merge with the protoplasmic network or unite with corresponding processes of adjacent cells, so that the cleavage cells, together with the other protoplasm of the egg—protoplasmic meshwork, cortical layer—form a single continuous whole, and in fact constitute a true syncitium.[3] As the cells move toward the surface they become more numerous by repeated divisions, and more closely crowded together, their connecting processes become correspondingly more numerous, shorter, stouter and more regular in outline. Figure 9 illustrates a tangential section through the layer of cleavage cells near the anterior end of an egg between Stages II and III showing the manner in which the cells are linked together at this time.

[3] See footnote p. 16.

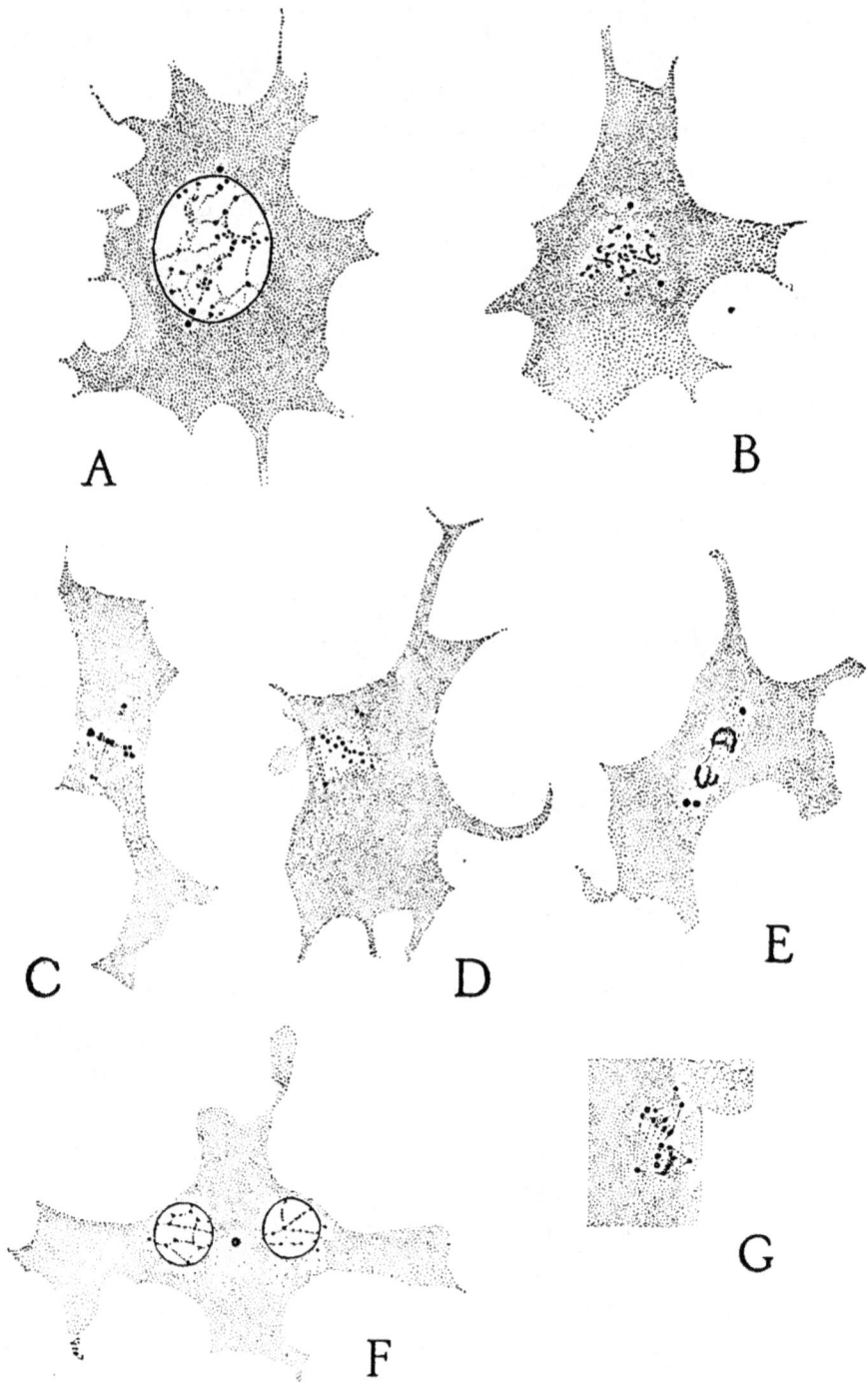

FIG. 8. Cleavage cells, showing stages of cell division. A and B, prophases; C and D, metaphases; E and F, telephases; G, multipolar spindle, x 1107.

The nuclei of the cleavage cells, during the resting stage, are circular to broad elliptical in outline. Each contains a network of achromatic material bearing on it a number of small rounded granules of chromatin. No nucleoli—karyosomes or plasmosomes—were observed. The size of the nuclei is, in a given egg, quite uniform, from the beginning to the end of the period under consideration, but varies considerably in different eggs, ranging from 9-14 microns.

The position of the nucleus within the cell is very variable, but it can be stated that in general the nuclei during the earlier Stages (I-II) occupy a position near the center of the cell body (Fig. 7B). As cleavage progresses, however, the nuclei evidence a tendency to take up a position near the margin, and this tendency becomes more and more marked as the cells approach the periphery of the egg (Figs. 6 and 7A).

Nachtsheim has recently (1913) published a detailed account of mitosis in the cleavage cells of both the fertilized and unfertilized eggs of the honey bee. Prior to the appearance of this account the writer had also studied this subject and prepared figures 8A-F, illustrating the different phases of mitosis. The writer's observations agree with those of Nachtsheim except in one or two minor details. A clearly defined spireme was never observed, although in the prophases (Figs. 8A and B) the chromosomes are seen to be united in short strings, which in many preparations, as in that from which figure 8B was drawn, give the effect of a relatively small number of curved rod-shaped chromosomes, and it is only by the use of the highest magnification available that these resolve themselves into the minute spheroidal chromosomes of which they are composed. No special effort was made to determine the number of these. It seems probable that Nachtsheim is right in estimating the number as thirty-two. There are certainly more than sixteen. The centrosomes,[4] densely staining minute spherical bodies, are readily seen at all stages. In figure 8A they have taken positions at opposite sides of the nucleus preparatory to division. Soon after this the nuclear membrane breaks down, usually disappearing first at the

---

[4] Nachtsheim, following Boveri, prefers to call these the centrioles, and the surrounding zone the centrosome. The latter the writer regards as the attraction-sphere.

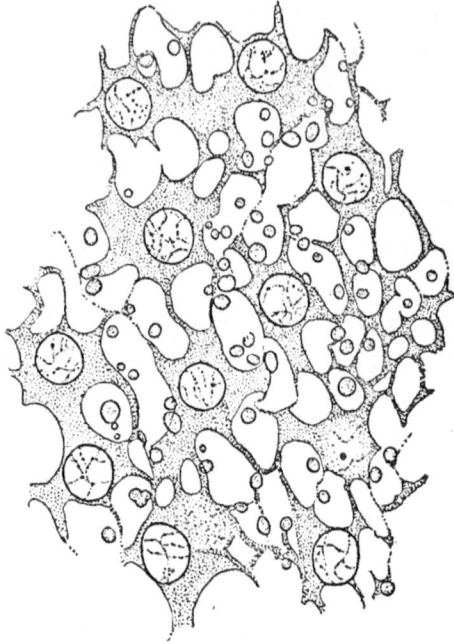

FIG. 9. Tangential section through cone of cleavage cells, Stage II-III, showing arrangement of cells and their relation to one another, x 534.

points opposite the centrosomes, when the astral rays may frequently be seen attached to the chromosomes and traversing the nucleus. These fibres are plainly evident in preparations fixed in Petrunkewitsch's fluid, but less so in those fixed in picro-formol, as were those from which the series of drawings were made. In figure 8B an attraction-sphere surrounds each centrosome. Figures 8C and D show the metaphases of division. One of the centrosomes is seen to have divided, and this division becomes even more evident in the anaphases and telophases (Figs. 8E and F). Nachtsheim describes a division of the centrosomes at both poles, which would naturally be expected, but the numerous preparations studied by the writer for some reason showed a division at only one pole. During the anaphases (Fig. 8E) the centrosomes increase in size and the nuclei frequently have a bilobed appearance. This division of the nuclei into two lobes is not uncommon in the early stages of development of other animal forms, as for example *Cyclops* (Rückert 1895) and *Crepidula* (Conklin 1897). One of the two lobes in such cases is supposed to represent the maternal half of the nucleus, the other the paternal, corresponding to the two pronuclei which unite to form the first cleavage nucleus.

A prominent and deeply stained mid-body is frequently observable between newly divided daughter nuclei (Fig. 8F).

A peculiarity of mitosis, confined to the cleavage cells, at once evident in glancing over the series represented in figs. 8A–H, is the small size of the spindle as compared with that of the resting nucleus. This disparity ceases to appear after the cells have attained the cortical layer and commenced to form blastoderm.

Multipolar spindles, as described by Lecaillon (1897a) were not observed during cleavage, but in certain preparations, irregular mitotic figures were visible of the type represented by figure 8G. These are so rare, however, that it would not be permissible to conclude that they are normal phenomena.

In a number of insect eggs,[5] it has been observed that the division of the cleavage cells take place in a definite direction, that is, at right angles to the egg's surface. Examination of sections of the bee's egg during Stages I and III, shows in dividing cells mitotic spindles turned in various directions and apparently not conforming to any rule, except when the cells are approaching the periphery of the egg. In this case the spindles are more frequently parallel to the surface, indicating that the resulting division plane will be normal to the surface. Nevertheless, a consideration of the relation of the cleavage cells to one another will make it evident that the majority of the divisions must be in effect at right angles to the egg's surface, since the cleavage cells—with the exception of those few destined to form yolk cells,—are arranged in a single layer, and this formation is maintained throughout the period of cleavage. The products of every division—with the exception just noted—must therefore eventually come to lie in this layer, so that whatever the directions of the mitotic spindle may be, the plane ultimately separating the daughter cells will be normal to the egg's surface.

In figures 5, 6, and 7B there is observable a marked contrast between the structure of the egg within the zone of cleavage cells and that outside of it. This difference concerns the protoplasmic meshwork. On the outside of the zone it is evident that the

[5] E.g., *Aphis, Musca,* Blochmann (1887); *Blatta,* Wheeler (1889); *Forficula,* Heymons (1895); *Clytra,* Lecaillon (1897a).

strands composing the network present the same appearance as
in the unsegmented egg; within the zone they are scarcely visible.
On examination of the interior region with high powers it is
seen that the network has now all but disappeared, of it there
remains only a cobweb-like remnant, with tiny polyhedral thickenings marking the nodes (Fig. 10). Since the network periphe-

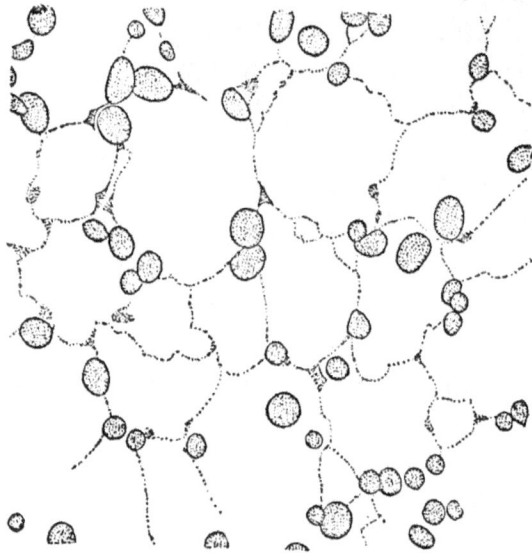

Fig. 10. Protoplasmic reticulum from center of egg, Stage II, showing
diminution of reticulum subsequent to the peripherad migration of cleavage cells, x 1107.

rad of the zone of cells is not altered, it is evident that what was
lost from the interior of the egg has been drawn into the
cleavage cells. This process of absorption continues as the cells
advance until they reach the cortical layer, when the condition
formerly seen only within the line of cleavage cells now prevails
throughout the entire egg, except in the immediate neighborhood
of the yolk cells. By far the greater portion of the protoplasm
originally present in the ovum is drawn to its surface to furnish
material for the building up of the future embryo. A similar
process of absorption of protoplasm by the cleavage cells doubtless exists in most insect eggs, but is much more strikingly evident in certain of the Hymenoptera than elsewhere. Carrière
and Bürger have described and figured it for both *Chalicodoma*
and *Anthophora;* in both of these forms it is identical with that
seen in the honey bee. Grassi and Dickel have also represented

in their figures the lack of protoplasm within the line of cleavage cells, without specifically mentioning it in the text.

Omitting an extended review of the results of other investigators who have dealt with this stage of insect development[6] it may be said that speaking broadly, the cleavage process in the bee is fairly typical for the higher insects, and offers but few peculiarities worthy of extended comment. Such differences as exist relate principally to details. Taking up some of these, it may be mentioned first, that while the formation by the cleavage cells of a hollow one-layered figure is common to most of the higher insects, yet this figure in most cases conforms more closely to the outline of the egg than in the case of the bee. This is illustrated by the Coleoptera (*Hydrophilus*, *Clytra*), Diptera (*Musca*), Lepidoptera (*Zygaena*) and some Hymenoptera (*Lasius*, *Polistes*). Second, that part on the egg's surface first reached by the cleavage cells may be located near the middle of the egg, or near one of the two poles. A table compiled by Marshall and Dernhehl (1906)[7] showing the location of this place in twenty-six insects belonging to seven different orders, also shows that different insects within the same order may differ widely in this respect. The Hymenoptera illustrate this point quite well, since, in *Formica* (Ganin 1869), *Lasius* (Henking 1892), and *Rhodites* (Weismann 1882), the cleavage cells first reach the periphery of the egg near the posterior end, in *Platygaster* (Kulagin 1897), near the equator, in *Chalicodoma* (Carrière and Bürger, 1897), and *Polistes* (Marshall and Dernhehl, 1906), near the anterior end, on the ventral side, as in *Apis*. It is therefore evident that this feature possesses but little significance.

In the ant *Camponotus*, according to Tanquary (1913) the conditions obtaining during cleavage are quite different from that of the majority of other non-parasitic Hymenoptera, and decidedly remarkable, recalling the conditions found by Weismann (1882) in *Rhodites* and *Biorhiza*. According to Tanquary's rather imperfect observations, a large nucleus, presum-

[6] For a complete review of this phase of insect embryology see Lecaillon 1897a, Henneguy 1904 and Marshall and Dernhehl 1906. The last named is particularly complete on the historical side.

[7] L. c. p. 140.

ably the first cleavage nucleus, is seen at the posterior end of the egg. Later smaller cleavage cells are found in the anterior half. These multiply, spread apart, and finally form the blastoderm, which, as in the honey bee, first appears on the ventral side of the egg, near the anterior end. The large nucleus at the posterior end of the egg meanwhile disappears.

In concluding it should be said that the account of the cleavage in the egg of the honey bee agrees essentially with those of previous investigators, as far as their observations extend. The accounts of Kowalevski, Bütschli and Grassi are, as regards the cleavage, rather fragmentary and incomplete; that of Dickel is much more complete and in general correct, although his figures are diagrammatic, and evidently not accurately drawn from actual preparations. Petrunkewitsch (1901) also gives a brief description of cleavage in the drone egg, illustrated by two figures. Among the other Hymenoptera investigated, those whose early development conforms most closely to the bee are *Chalicodoma* and *Polistes*.

IV

Formation and Completion of the Blastoderm

In the preceding section the cleavage cells were followed to the point where they reach the periphery. This stage in their migration is illustrated by figures 7A and 11A. In the second figure mentioned only a single cell is represented, much magnified, together with a corresponding portion of the cortical layer. A distinction is now apparent between the protoplasm of the cortical layer and that of the cleavage cells. The former is clear and transparent, the latter dark and granular. The cleavage cells rest upon the cortical layer as upon a base, supported by short intervening processes which are separated from one another by irregularly circular openings. On meeting the cortical layer these processes spread out to unite under this layer to form a continuous sheet. On comparing figure 7A with an earlier stage, figure 6, it becomes evident that these processes extending peripherad from the cleavage cells are only the strands of the protoplasmic meshwork underlying the cortical layer, shortened and thickened, while the openings represent the spaces occupied by the vitelline spheres. As the cells advance peripherad the processes shorten, while the open spaces flatten and ultimately disappear (Fig. 11B) ; the nucleus of each cell is thus brought into contact with the cortical layer (Fig. 11C). The nucleus continues to advance peripherad, followed by the deep staining protoplasm of the cell body, while over it the cortical layer begins to rise up on the exterior in a more or less rounded swelling, each of these being marked off from its neighbors by a wide and deep furrow (Figs. 11C, D and E). This stage, when seen in the living egg, presents a very striking and characteristic appearance, and has been well described and figured by Kowalevski (1871). A differentiation in the character of the protoplasm of the cell body now begins to appear. That part accompanying the nucleus is seen to be less deeply stained and

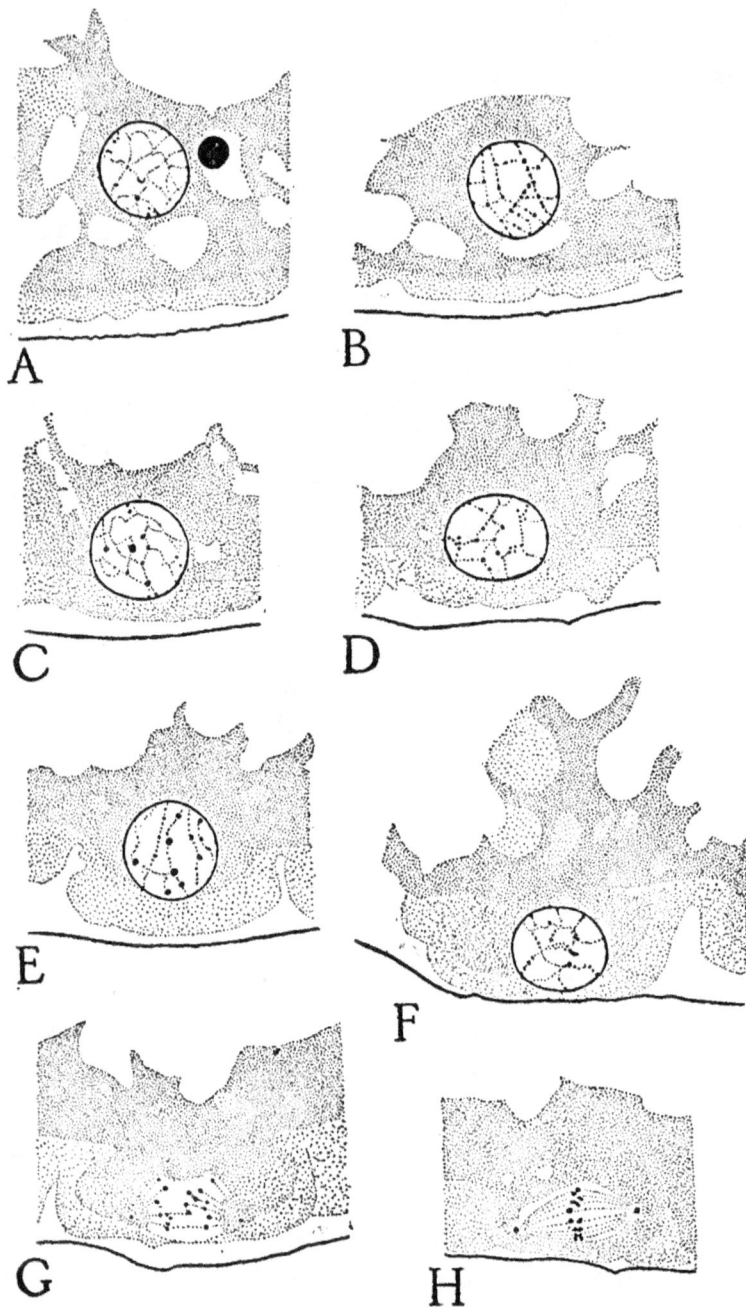

Fig. 11. A series of stages to illustrate the entrance of a cleavage cell into the cortical layer during the formation of the blastoderm. In A the cleavage cell has just come into contact with the lighter tinted cortical layer. In F the external portion of the cleavage cell has entered the cortical layer and has produced a well marked elevation on the surface of the egg. In G and H the cleavage nucleus is dividing, x 1107.

less granular than the deeper lying portions, although it, in turn, is still less transparent and darker than the protoplasm of which the cortical layer is composed, and sharply marked off from it. In figure 11F the migration is nearly completed; each cleavage nucleus is embedded in a rounded swelling of the cortical layer, and flanked on each side by a mass of cytoplasm derived from the cleavage cell; these are united behind (centrad to) the nucleus and merge with a still darker mass of cytoplasm which still retains the branching and amoeboid form characteristic of the cleavage cells. These irregular masses of dark cytoplasm belonging to adjacent cells now form a continuous although irregular stratum underlying the nuclei with their accompanying cytoplasm. Next the cell boundaries become distinguishable, faint indications of cell walls extending inwards from the bottom of the furrows; the deeper stained bases of the blastoderm cells are however still linked to one another, so that cell territories are yet not completely demarcated (Fig. 12A). Soon after, the walls separating the cells are completed; a delicate basement membrane is formed near the bases of the cells, cutting off the latter from the narrow inner zone of deeply staining protoplasm. The cells are now completely delimited and the blastoderm, as such, is established (Fig. 13A). It should, however, be always borne in mind that these processes do not take place simultaneously over the entire surface of the egg, or even over any extended part of it. They commence where the cleavage cells first reach the cortical layer, namely, on the ventral side near the anterior end, then extend rapidly over the entire anterior third of the egg, thence progressing slowly caudad, so that all phases of the process of blastoderm formation may frequently be observed in a single egg.

A striking feature of these early phases in the formation of the blastoderm is the rôle played by the nucleus. As mentioned in the section preceding, when the cleavage cells approach the cortical layer, the nuclei take up a peripheral position in the cells. This is plainly seen in figures 7A and 11A; the nuclei are now very close to the external margin of the cell body. From this point on the nucleus leads the advance into the cortical layer, and after coming into contact with it (Fig. 11B) has apparently become denuded on its peripheral side of cytoplasm belonging

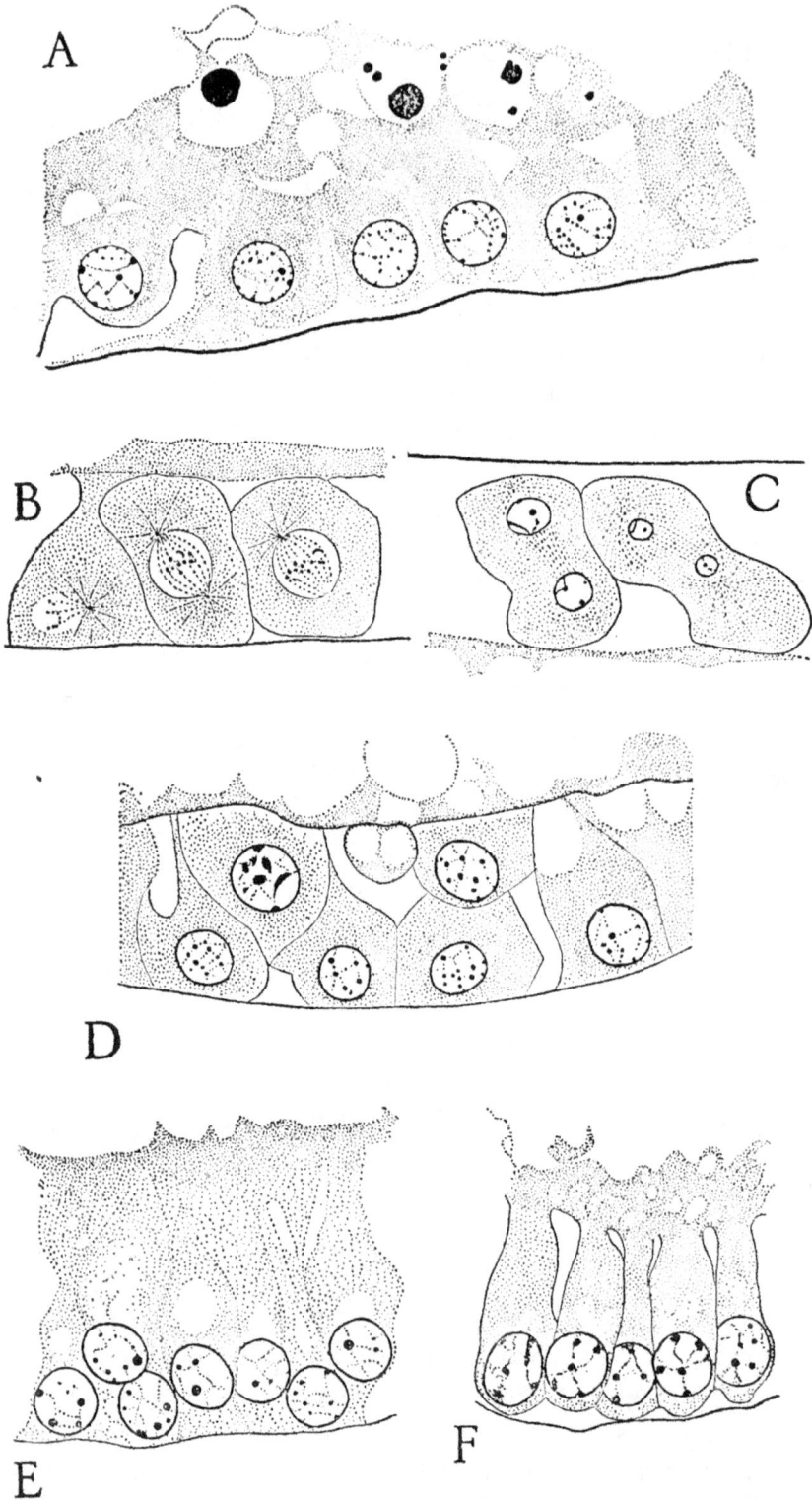

FIG. 12. Later stages in the formation of the blastoderm. A, newly formed blastoderm, from a sagittal section, cell boundaries just appearing. B and C, blastoderm cells dividing obliquely. D, blastoderm cells so arranged that their nuclei lie at two levels. E, blastoderm cells from ventral side of egg, 24-26 hours. F, same, 28-30 hours, x 1107.

to the body of the cleavage cell. That this migration is accomplished by an actual movement on the part of the nucleus itself is contrary to the facts of biology; the source of the movement must therefore be referred to the cytoplasm, as Marshall and Dernhehl have pointed out. Moreover, nuclei are frequently seen elongated at right angles to the line of movement, as though somewhat flattened by pressure before and behind (Figs. 11D and F). A flattening of the nuclei has also been observed to occur under similar circumstances by Lecaillon (1897a) in *Clytra*, and by Noack (1901) in *Musca*, and is similarly explained by these investigators. In these instances, however, the nuclei are flattened only at their peripheral ends.

A second feature to be noted is the sharp separation of the protoplasm derived from the cell body and that of the cortical layer. This separateness is readily seen in figure 12A, when the cell walls are beginning to appear. An ultimate fusion of the material from these two sources occurs later, but seems to take place only very slowly. The cytoplasm of the blastoderm cells is therefore derived in part from the cleavage cells, but not all of the cytoplasm from this source is thus employed, a considerable portion being temporarily cut off and left within the blastoderm in the form of a deeply stained granular layer, described above. Such a layer of granular protoplasm was first observed by Weismann (1863) in *Musca* and named by him the "innere Keimhautblastem"; this layer may then accordingly be termed *the inner cortical layer*, but it must be remembered that it has no close genetic relationship with the cortical layer proper. This inner layer has subsequently been found to exist in the eggs of many insects, but appears to be wanting in the orders Orthoptera and Dermaptera.

Figures 13A and 13B represent transverse sections through an egg just after the blastoderm has become established. As is readily apparent, the blastoderm is not of uniform thickness throughout, but is thinner on the dorsal than on the ventral side. Comparing figure 13B, from the posterior region of the egg, with figure 13A, through its anterior region, it is evident that the blastoderm-forming cells of the posterior region are much less numerous than those of the anterior region. This difference is directly referable to the preceding cleavage stages when a

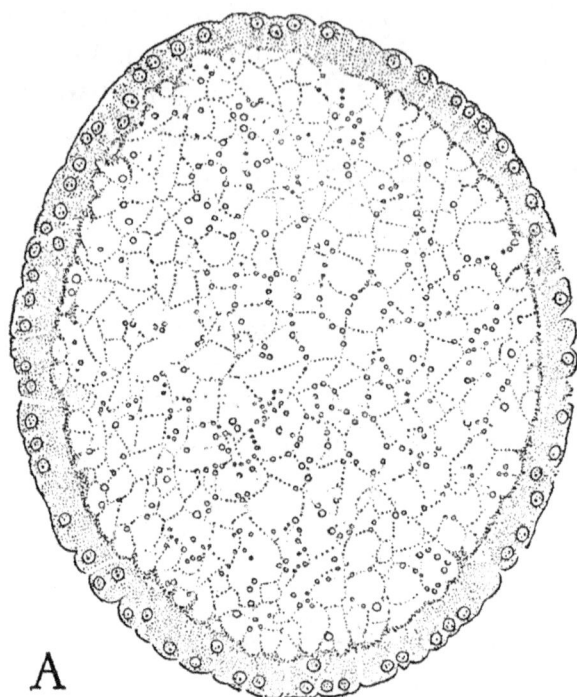

A

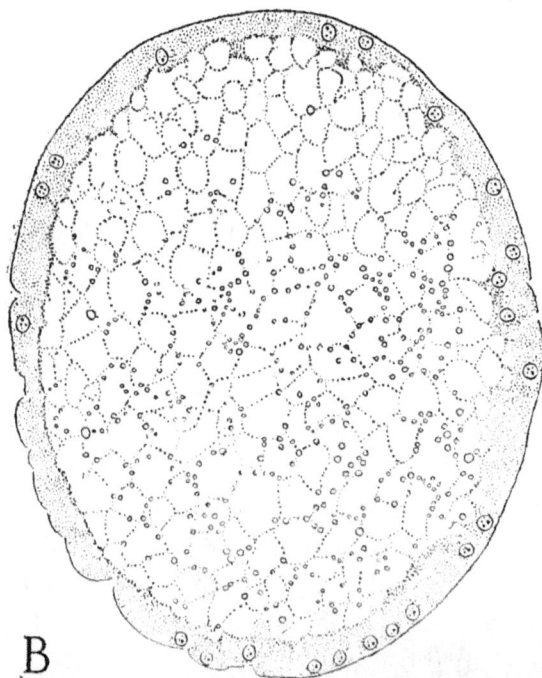

B

FIG. 13. Tranverse sections from egg about 16 hours old, just after formation of blastoderm. A, near anterior, B, near posterior end, x 243.

similar numerical relation occurs (see p. 16 and Figs. II and III).

Figure 14 represents a sagittal section of a stage a trifle older

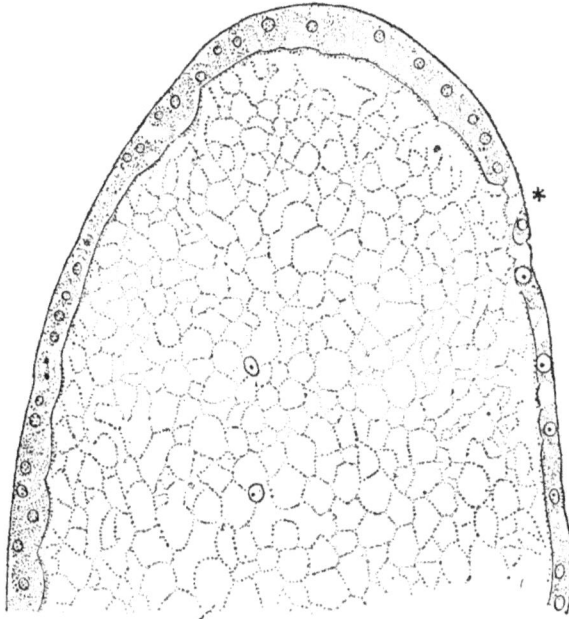

FIG. 14. Anterior end of a sagittal section of an egg about 16 hours old. The point of junction of the future ventral blastoderm with the thin dorsal blastoderm is designated by a star, x 243.

than those of the two figures preceding. Here the difference in thickness between the dorsal and ventral blastoderm is more marked. Moreover the thicker blastoderm of the ventral side is continued over the anterior pole, being at this point even thicker than the ventral blastoderm itself. At the point on the dorsal surface where the thickened area joins the thinner and somewhat flattened cells of the mid-dorsal area (marked in the figure by a star) the cells of the latter have an irregular amoeboid form, and separated by short intervals, as though pulled apart. The existence of this area of scattered cells was first noticed in the bee by Dickel (1904) who has made much of it in his endeavor to construct an original interpretation of the origin of the germ layers in insects. A discussion of Dickel's view of the relation of this area to the germ layers will be given later.

About the time that the nuclei enter the germ layers they undergo mitotic division (Fig. 11G and H). This may occur either during the migration (Fig. 11H) or immediately after (Fig.

11G); the division plane is always normal to the external surface.

In many accounts of the insect development some details are given of the manner in which the cleavage cells became transformed into blastoderm, but probably the most circumstantial accounts are those of Blochmann (1887a) and Noack (1901) for *Musca;* of Heider (1889) for *Hydrophilus*, Carrière and Bürger (1897) for the mason bee (*Chalicodoma*) and its parasite, *Anthophora*, and Marshall and Dernhehl (1905) for *Polistes*. The last named account is by far the richest in details. In *Musca* and *Hydrophilus* the process (as described) is essentially the same. The cleavage cells, on arriving at the cortical layer, fuse with it to form a continuous sheet without any cell boundaries—a syncitium, in other words—in which the nuclei divide, the spindles of the mitotic figures being parallel to the egg's surface. Cell boundaries now appear, first as furrows of the external surface of the egg, thence extending inward, delimiting the cells on their lateral surfaces. Meanwhile the darker staining bases of the cells form a continuous layer within, constituting the "inner Keimhautblastem." In *Hydrophilus* this is soon absorbed by the cells of the blastoderm. In *Musca* this layer is much thicker and, in certain portions of the egg, separated from the blastoderm by a layer of yolk. Moreover it is only slowly absorbed by the blastoderm cells, but is ultimately taken up by them. In *Anthophora* the process is as just described, except that no "innere Keimhautblastem" is formed. It is also absent in *Chalicodoma*, as is the cortical layer (Keimhautblastem). The account of Marshall and Dernhehl allows a much closer comparison with the condition met with in the honey bee. In *Polistes* the cleavage cells enter into the cortical layer separately, with the nuclei situated at the peripheral margin of the cell body, which has the appearance of trailing after the nucleus like the tail of a comet, remaining meanwhile separate from the cortical layer and distinguished by its darker shade, as in the honey bee. A regular inner cortical layer was not observed, but subsequent to the formation of the blastoderm irregular patches of a rather granular mass were found at the bases of some of the blastoderm cells. The separation of the cell territories was accomplished in the usual manner. There is then almost complete agreement between *Polistes* and *Apis* in the particulars relating to the entrance of the cleavage cells into the cortical layer and their subsequent transformation into blastoderm. The chief points of difference are: the narrow, almost pointed shape of the cleavage cells of *Polistes* when entering the cortical layer, and the relatively scanty amount of basal (central) material left to form an inner cortical layer.

The further changes in the history of the blastoderm can best be followed by taking them up in the order of their age. This is in fact the only safe method of determining the order of development during this period.

Seventeen to twenty hours. The formation of the blastoderm, it is estimated, is completed in from 14 to 16 hours, since the stage under consideration, identified by means of eggs whose age is actually known, immediately follows and is directly connected with the stages just described by changes which can occupy only a brief interval.

While the cleavage cells are entering the cortical layer, or immediately after this event, they divide mitotically, with the spindles parallel to the egg's surface, as already described. This may possibly be followed by another similar division, but whether this occurs is uncertain; at all events not more than two divisions of this character intervene between the entrance of the cleavage cells into the cortical layer and the stage of 17-20 hours. Subsequent mitotic divisions however do occur, probably in every blastoderm cell, but the spindles of the mitotic figures are no longer uniformly parallel to the surface of the blastoderm; they now more frequently are inclined at an angle to it, in other words, they are oblique (Fig. 12B and C). As these illustrations show, the angle formed by the spindles of different cells with the surface of the blastoderm is not uniform; some spindles are parallel to the surface, as in the left hand cell of figure 12B; some form an angle of 45 degrees with it, as in the middle cell of the same figure, while others are inclined at a lesser angle. The effect of these oblique divisions is shown in figures 12D and 15. At first glance the blastoderm appears now to consist of two layers of cells, superimposed one above the other, since two rows of nuclei are seen, except along the dorsal mid-line. That two layers of cells are actually present is rendered improbable

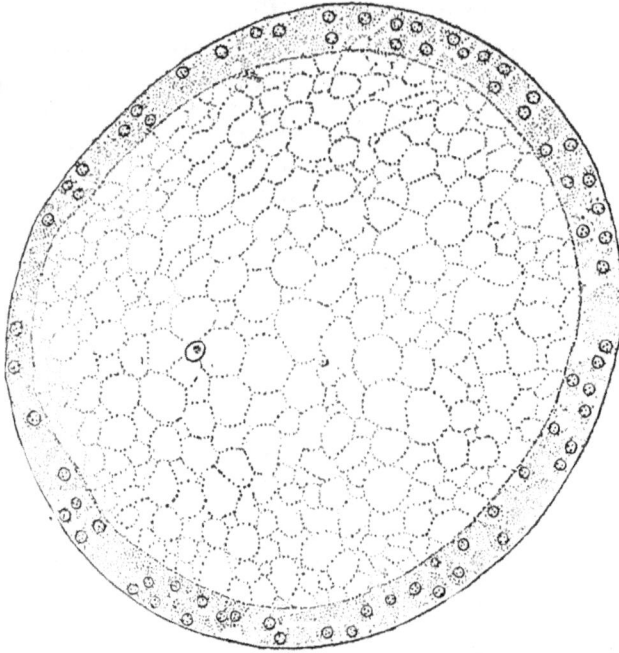

Fig. 15. Transverse section through the anterior region of an egg about 17-20 hours old, showing the nuclei of the blastoderm arranged in two layers, x 243.

by the fact that very many of the cells situated next to the peripheral surface are pyriform in outline, with their slender ends reaching inwards to the basement membrane. These alternate more or less regularly with shorter cells lying close to the basement membrane, but which frequently possess a slender peripheral process. This apparent two layered condition was described and figured by Kowalevski (1871), who explained it ingeniously by supposing it to be an artifact, brought about during the formation of the action of the fixing agent. The latter was assumed to have contracted the layer of cells first appearing at the surface of the egg, closing up the intervals between them and crowding them down on those cells arriving later. This same stage has also been figured for the drone egg by Petrunkewitsch (1901, Fig. 24), a dividing cell is also shown. Oblique divisions of the blastoderm cells are also described for *Polistes* by Marshall and Dernhehl (1905).

Viewing the blastoderm as a whole, it appears more compact and sharply outlined than it was directly after its inception; it now constitutes a well developed epithelium. Both its outer and inner surfaces are regular and smooth in contour. On the inner surface the inner cortical layer is evident, but this is also more uniform as well as thinner than before. Moreover the processes on its inner surface are no longer so strikingly apparent, although they still exist.

Twenty-four to twenty-six hours. This stage is represented by a sagittal section, figure 16. The blastoderm at this stage

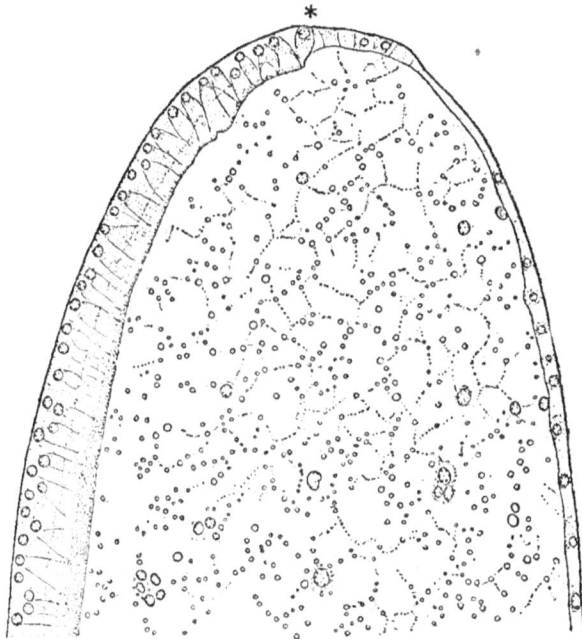

Fig. 16. Anterior portion of a sagittal section of an egg about 24-26 hours old. The contrast between the thick ventral blastoderm (ventral plate) and the thin dorsal strip is very marked. The point of junction between them is marked by a star, x 243.

differs from that of the preceding stage in several particulars. The contrast between the blastoderm on the dorsal surface and that over the remainder of the egg is very marked, since there is now evident, along the dorsal mid-line, a fairly well developed strip about one-third of the diameter of the egg in width, and extending from one pole to the other, composed of flattened cells. On the other hand, the cells of the ventral and lateral surfaces have become elongated and columnar in form, whereby, the thickness of the blastoderm in these regions is much increased. In addition the thickened blastoderm over the anterior pole has disappeared, giving way to a thin sheet of cytoplasm containing scattered nuclei; the boundary between the thickened blastoderm of the ventral surface and the thin dorsal sheet is sharp, and is situated precisely at the anterior pole (marked by a star), instead of dorsad of it, as in earlier stages (*cf.* Figs. 14 and 16). This change is apparently brought about by a shortening of the ventral blastoderm in the longitudinal axis, since stages intermediate in age (22-24 hours) show intermediate conditions in respect to the position of the point of juncture of the thicker ventral blastoderm and the thinner dorsal strip. In addition to these differentiations of the blastoderm, its average thickness is much greater in the anterior than in the posterior half of the egg. Beginning at the cephalic pole and passing backward the blastoderm first rapidly increases in thickness as far as the middle of its anterior half (Fig. 16), from this point it grows thinner slowly but uniformly to the posterior pole. A slight difference in thickness in favor of the anterior half of the egg is observable at all stages, but at the present one is especially noticeable.

The cells composing the blastoderm in the available preparations of this stage present on close examination a very peculiar aspect. All—except those of the dorsal strip—have now elongated to a somewhat irregularly prismatic form, with their bases frequently constricted. The nuclei are situated close to their peripheral ends (Fig. 12E). The cells have not however all elongated to the same extent, so that the nuclei do not yet lie uniformly at the same level. The cytoplasm of the cells is pale, and finely vacuolated in structure; centrad of each nucleus is a large vacuole approximating the nucleus in size. The lateral cell walls are seen to pass inward some two-thirds of the thickness of the blastoderm, where they are lost to sight in a zone of vacuolated cytoplasm formed by the bases of the cells themselves, which have become continuously fused with one another. Centrad of this zone, but merging with it, is a darker one, recognizable as the inner cortical layer (*ICL*). Referring to figure 12A, it will be readily apparent that the blastoderm cells, so far as their relation to one another and to the inner cortical layer is concerned, have resumed the condition existing at the time that the blastoderm was first established, the basement membrane formerly underlying the blastoderm cells and separating them from the inner cortical layer having disappeared. A discussion of the significance of this change will be deferred to the end of the section.

Twenty-eight to thirty hours. At this stage (Fig. 17) a cross

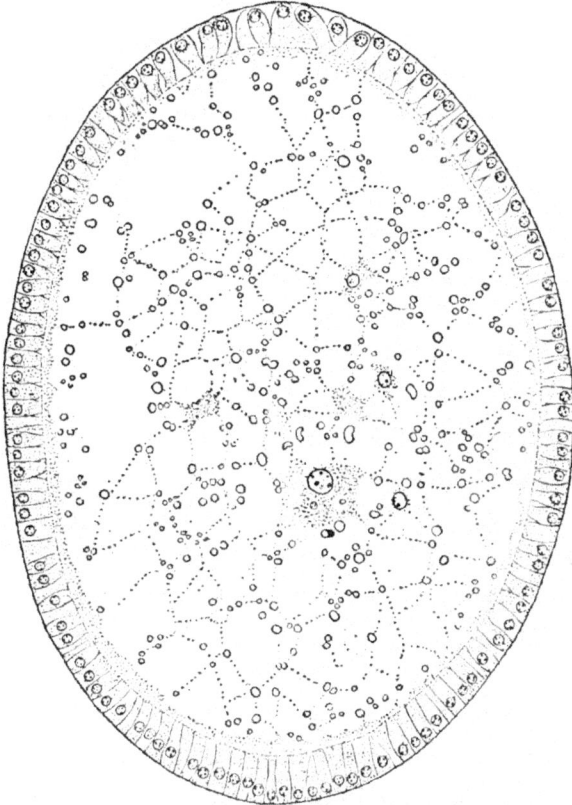

Fig. 17. Transverse section through the anterior half of an egg about 28·30 hours old. All of the cells of the blastoderm, including those of the dorsal side, have assumed a columnar form, x 243.

section through any portion of the egg shows the blastoderm of nearly equal thickness over the entire circumference, the dorsal thin strip of the stage preceding having regained a thickness equivalent to the remainder of the blastoderm; its cells neverthe-less are distinguishable from those of the remainder of the blasto-derm by their irregular form. In general the blastoderm cells are regularly prismatic, much shorter and more compact in structure than in the preceding stage, and slightly constricted near their bases (Fig. 12F). The latter are, as before, fused to form a continuous layer, but this is now thin, dark and granular, ap-parently composed of material derived from the inner cortical layer. Each cell still possesses an elliptical vacuole lying centrad

of the nucleus, in those cells situated on the dorso-lateral faces
of the egg the vacuoles are larger and much more prominent.

Thirty-two to thirty-four hours. (Fig. 18A and B.) This
stage is distinguished from the preceding principally by two
points of difference: first, the blastoderm cells now form a
columnar epithelium, they are no longer joined by their bases
but limited at their central ends by a well defined basement membrane; second, the mid-dorsal area has again become reduced to
a thin sheet of flattened cells. Its point of juncture at the anterior end with the thick blastoderm of the ventral side has now
become shifted to a point ventrad of the cephalic pole (Fig. 18A).
The ventral blastoderm has therefore undergone a further shortening in the long axis of the egg. Of the inner cortical layer
there remains a vestige in the form of a very thin granular layer
lying near the inner side of the blastoderm.

Since this stage closes what may be called the blastodermal
period of the development, and brings the account down to the
formation of the germ layers—which is in fact inaugurated during this stage—a brief discussion of some of the phenomena
characterising this period is in order.

Among other features, the behavior of the blastoderm cells
toward the inner cortical zone deserves especial mention. To
review this process briefly: the cleavage nuclei enter the (outer)
cortical layer accompanied by a portion of the cytoplasm belonging to the cleavage cells, but leaving another darker portion
within to form a basal layer, the inner cortical layer, by which
the newly formed blastoderm cells are at first united to each
other. This is next cut off by the formation of a basement membrane. After repeated divisions the cells elongate, the basement
membrane disappears and the inner cortical layer is taken up by
the cells (22-26 hours). A new basement membrane is then
formed (32-34 hours), when the cells have the arrangement and
appearance of ordinary columnar (prismatic) epithelial cells.
Turning again to the accounts of the development of *Hydrophilus* (Heider 1889) and *Musca* (Noack 1901), similar phenomena are found to exist in these forms. In *Hydrophilus* the
inner cortical layer is taken up into the blastoderm cells in precisely the same manner as in the honey bee. Moreover the blastoderm cells afterwards present the same elongated form and con-

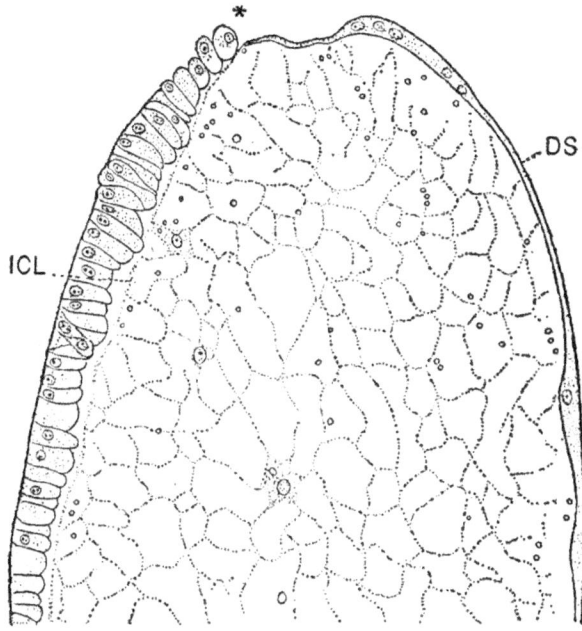

A

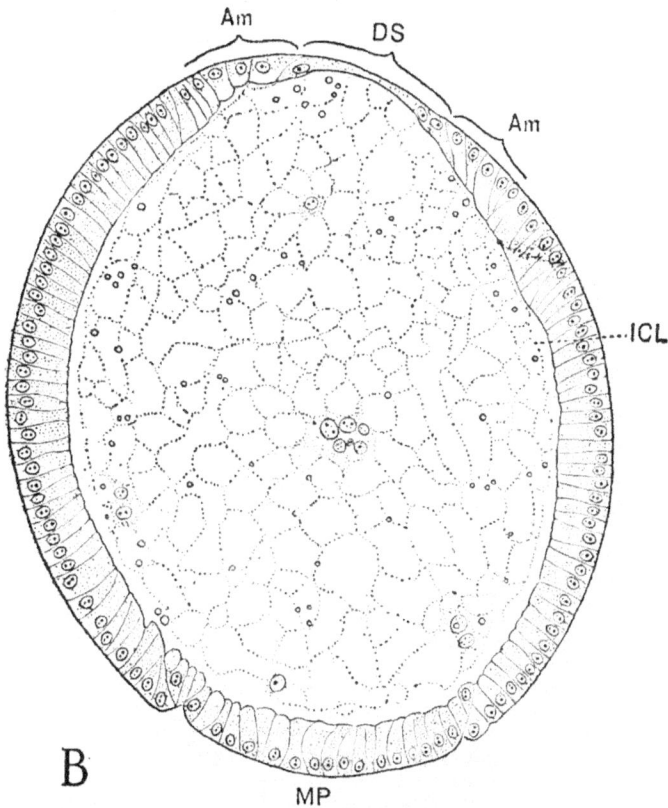

B

FIG. 18. Sections of eggs about 32-34 hours old. A, Anterior part of median sagittal section. The point of junction of the ventral blastoderm with the thin dorsal strip s marked with a star. B, Transverse section near anterior end. The thin dorsal strip (DS), the cells destined to form the amnion (Am) are plainly shown. The middle plate (MP) is also becoming differentiated, x 243.

stricted base. There is however, this difference: that in *Hydrophilus* the blastoderm cells maintain continuously their connection with the inner cortical layer, and are never separated from it by a basement membrane. In *Musca,* however, in certain parts of the egg, the inner cortical layer is separated from the blastoderm cells by a layer of yolk granules. By the inward (centrad) extension of the blastoderm cells, this layer of yolk, as well as the inner cortical layer, is taken up into the bases of the cells. This process seems to take place much more slowly and at a later period than in *Hydrophilus*. In this respect, as well as in the separation of the inner cortical layer from the blastoderm cells, *Musca* may be said to approach the honey bee more closely than *Hydrophilus*.

Another feature of interest is the long duration of the blastodermal period—16 to 18 hours, approximately, or nearly one quarter of the total length of time required for complete development in the egg. In *Hydrophilus* (Heider 1889), the only insect concerning which full and accurate data are given of the duration of each stage, the period required for the formation and completion of the blastoderm (from the close of cleavage to the first appearance of the lateral plates) is about 27 hours. The total time required for development in the egg is 11 days, or 274 hours. The blastodermal period then occupies a trifle more than one-tenth of the entire time, as compared with one-fourth in the honey bee. This comparison, however, is perhaps a little unfair, since the larva of *Hydrophilus* at hatching is fitted for an active existence and probably therefore more highly differentiated than the larva of the bee at a corresponding stage; nevertheless, the comparison is a striking one.

The significance of the long sojourn of the bee's egg in the blastodermal period can only be guessed at; its importance can only be estimated by its duration. Compared with the changes occurring within the egg in any other equal space of time the morphological changes during this period are insignificant. This leads naturally to the surmise that the nature of the changes undergone by the egg at this time may be principally physiological, and therefore not made evident by the ordinary methods of the embryologist.

V

The Germ Layers

1. *Formation of the mesoderm*

At the time when the germ layers begin to be formed, (32-34 hours) the blastoderm presents three more or less distinct divisions: (1) a median dorsal strip, (2) the blastoderm of the ventral and lateral surfaces, (3) two bands, one on each side of the dorsal strip (1), between the latter and (2).

The median dorsal strip (Fig. 18A, *DS*) extends to the caudal pole, from a point slightly ventral of the cephalic pole. It is widest near the cephalic pole, attaining here a width of about one-third of the circumference of the egg in this region, caudad of this point it narrows gradually and over the remaining distance continues quite narrow (Fig. 18B). It is composed of a relatively small number of extremely thin and flat cells, and also differs from the remainder of the blastoderm in its close relation to the protoplasm surrounding and permeating the yolk. As may be readily determined in most preparations the yolk is enclosed in a pellicle of protoplasm, derived from the inner cortical layer, forming a part of the interstitial protoplasmic meshwork and continuous with it. This pellicle is commonly separated from the basement membrane of the blastoderm by a narrow space. This relation, however, ceases at the lateral edges of the dorsal strip, since here the protoplasmic pellicle unites and becomes continuous with the lateral edges of the cells composing the strip. This connection becomes especially evident during the formation of the germ layers and the amnion. As a result of the disappearance of the pellicle under the dorsal strip the latter supplies the place of the pellicle and thus comes into close relation with the yolk and the meshwork of protoplasm within it. The fate of the dorsal strip will be dealt with later, but it may be said in advance that it takes part neither in the formation of the embryo nor its covering (amnion), and is therefore strictly non-embryonic.

43

The blastoderm covering the ventral and lateral faces of the egg, as already described (p. 40), is a thick single layered epithelium composed of slender prismatic cells. Its average thickness is greatest in the anterior region of the egg and decreases slightly and gradually toward the caudal pole. This division, composing the major portion of the blastoderm, is the distinctly embryonic portion, since it alone contributes directly to the formation of the embryo.

On each side of the dorsal strip is a band comprising some four to six longitudinal rows of cells whose central ends are very clear and transparent (Fig. 19, *Am*). Those cells next to the

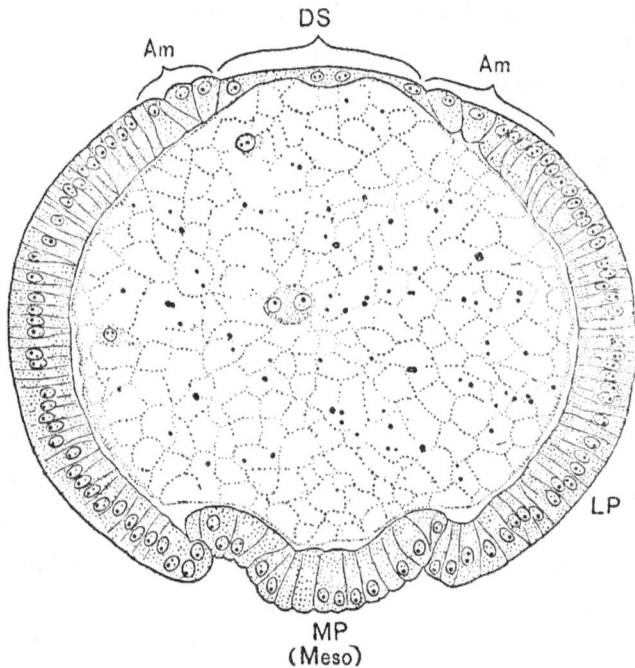

FIG. 19. Transverse section through the anterior region of an egg, Stage IV, showing dorsal strip of blastoderm (*DS*), amnion forming cells (*Am*), middle plate (*MP*) and lateral plates (*LP*), x 243.

dorsal strip are triangular in outline when seen in transverse section, and in contact with the dorsal strip at one corner; all the cells of this strip are less slender than the cells of the embryonic blastoderm (2), many of them approaching the cubical form. Those lying nearest the embryonic blastoderm intergrade with the cells of the latter, so that the lateral limits of this division of the blastoderm cannot be determined with precision until the formation of the germ layers is well under way. These two bands, which will later form the amnion together with the strictly embryonic division, may be conveniently considered as together constituting the *ventral plate*. Later, when the amnion begins to be more clearly differentiated, the embryonic portion of the ventral plate, which now constitutes the embryonic rudiment, is commonly known as the *germ band*.

This stage is introduced by the appearance, on the ventral surface of the blastoderm, of two narrow longitudinal ridges, sharply defined on their inner margins, less so on the outer. As seen in surface view, they appear as shown in figure IV, where they have the appearance of gently curved and slightly irregular dark lines with their convex sides facing one another, like reversed parentheses,)(, and separated from each other at the point of closest approximation by a distance of about one-eighth of the total circumference of the egg at this point. The distance between their anterior ends and the cephalic pole of the egg, is about equivalent to the egg's diameter here; their length when first visible from the exterior is, as shown in the figure, about one-fourth of the total length of the egg. As development proceeds these ridges extend rapidly caudad, diverging slightly at first and then pursuing a course nearly parallel to one another as far as the caudal pole of the egg, embracing between them a strip of blastoderm whose average width is almost one-sixth of the circumference of the egg at any given point (Fig. V). This strip is the *middle plate,* and constitutes the future mesoderm, while the embryonic blastoderm laterad of it on each side forms the *lateral plates,* whose mesial edges are the *lateral folds.*

Simultaneous with the caudad extension of the lateral folds (Stages IV-V) is a movement of their anterior portions toward the mid-line. This movement begins first at about that point where the folds originally approached one another most closely, that is, about one-fourth of the length of the egg from its cephalic pole. The anterior ends of the folds meanwhile remain stationary. The outline of the folds in consequence of their movement toward one another near their anterior ends, forms a figure more or less resembling an elongated flask (Fig. V). In the corresponding stages of *Hydrophilus* (Kowalevski, 1871), and *Chalicodoma* (Carrière and Bürger, 1897), the resemblance between the outline of the folds and that of a flask also occurs and is much closer. The lateral folds continue to approach one another until they meet, their first point of juncture being naturally that of most rapid movement, a short distance caudad of their anterior ends, which up to this time remain nearly stationary. The process of meeting and fusion now progresses both forward and backward from the first point of juncture, the anterior portions of the lateral plates being very quickly united, while the posterior come together rather slowly. Figures V and VI illustrate two stages in the completion of the process of closing; in the former figure a narrow cleft extending nearly one-half of the length of the germ band separates the lateral plates; in the latter figure this cleft is insignificant. With the final closure of this cleft the formation of the germ layers may be considered complete.

Examination of sections of the stages described shows that the process just described consists essentially in the depression of a median area of the ventral plate—the *middle plate,*—and its overgrowth by the lateral portions of the ventral plate—the *lateral plates,*—which have broken away from its edges along the line indicated by the lateral folds at the time of their first appearance. In respect to the manner of formation of the mesoderm the observations recorded above are in complete agreement with those of Kowalevski (1871) and Grassi (1884). As already mentioned (p. 40) the beginning of the process of mesoderm formation ("gastrulation," so-called) is indicated in figure 18B. In this figure a median section of the ventral blastoderm is seen to be in process of separation from the lateral portions by the mere displacement of the cells on each side of the boundary line. That is, the cells forming the margin of the nascent middle plate appear to be sliding inward over those forming the edges of the future lateral places. This stage is the earliest observed. Figure 19 represents a transverse section through the anterior end of an egg at Stage IV. Here the middle plate has become still more depressed, while the edge of the lateral plate on the left side of the figure has already overlapped the corrsponding edge of the lateral plate; on the right side of the figure the relation between the middle and lateral plates is much the same as in figure 18B. In sections like that represented by figure 19 it is evident that the middle plate has been actually depressed, especially at its lateral margins, as the corresponding deformation of the yolk shows, while the lateral plates are not perceptibly lifted up from the yolk. Figures 20, A and B, illustrate the stage next following. Figure 20A is a cross section through the egg near the point when the lateral folds are closest together; figure 20B is near their posterior ends. The middle plate has at this stage changed somewhat in structure in its anterior portions. The marginal cells, formerly like the remainder of the ventral blastoderm, slender and prismatic, with their long axes directed radially, have turned laterad under the lateral plates, becoming irregularly polyhedral in form. Corresponding with this change the middle plate becomes much broader and thinner, especially at its edges. Figure 20B illustrates the formation of the middle plate in the posterior half of the egg. It is evident at a glance that the process is identical with that in the more anterior regions of the egg. There is, however, this difference: that before the middle plate shows any signs of separation from the lateral plates it is distinguishable by being thinner than the lateral portions of the blastoderm, and its cells are correspondingly shortened and widened. This is evident in figure 20B. It is also evident that the initial width of the middle plate is greater in the posterior than in the anterior regions of the egg. This is probably correlated with the thinning of the incipient ventral plate which brings about a corresponding extension of its surface.

The final stages in the formation of the mesoderm are shown in figure 21, which is a transverse section through the middle of an egg a trifle younger than Stage VI, showing the lateral folds about to unite in the mid-line, the middle plate being now virtually completely covered. Its width is—at this point—fully equivalent to one quarter of the circumference of the egg; near the ventral mid-line its thickness is approximately that of the overlying lateral plates (ectoderm), it thins out rapidly to one-half of this thickness at the lateral margins. Here it is composed of but one cell layer; near the ventral mid-line it can be distinctly seen to be made up of two layers. This arrangement of the mesoderm cells into two layers is more evident in sagittal sections passing somewhat laterad of the median plane (Fig. 23B). It is also more evident in the posterior than in the anterior half

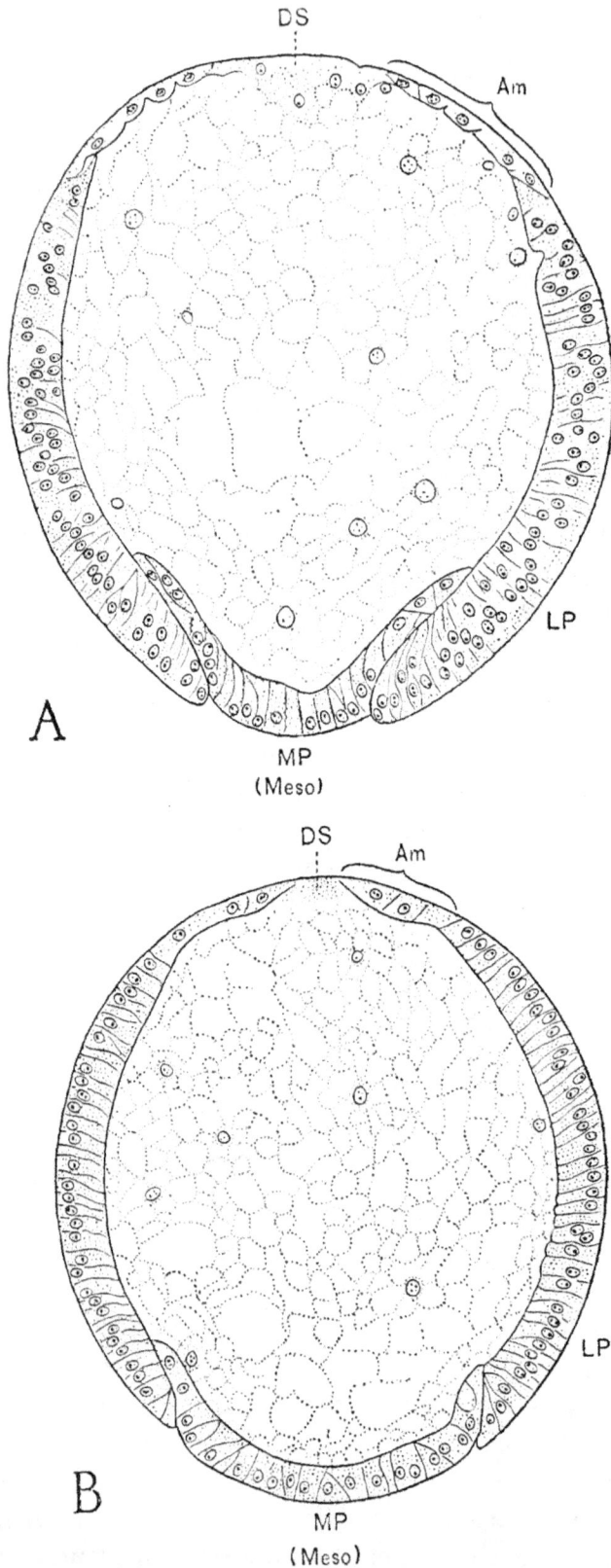

FIG. 20. Tranverse sections through egg, Stage V. A, near anterior, B, near posterior end. The middle place (*MP*), lateral plates (*LP*), amnion-forming cells (*Am*) and dorsal strip are shown, x 243.

of the mesoderm. In the rearrangement of the cells which are to form the mesoderm, cell division plays an important part, since cells in division are very abundant in the middle plate during Stages V and VI.

In the formation of the mesoderm *Apis* differs somewhat from

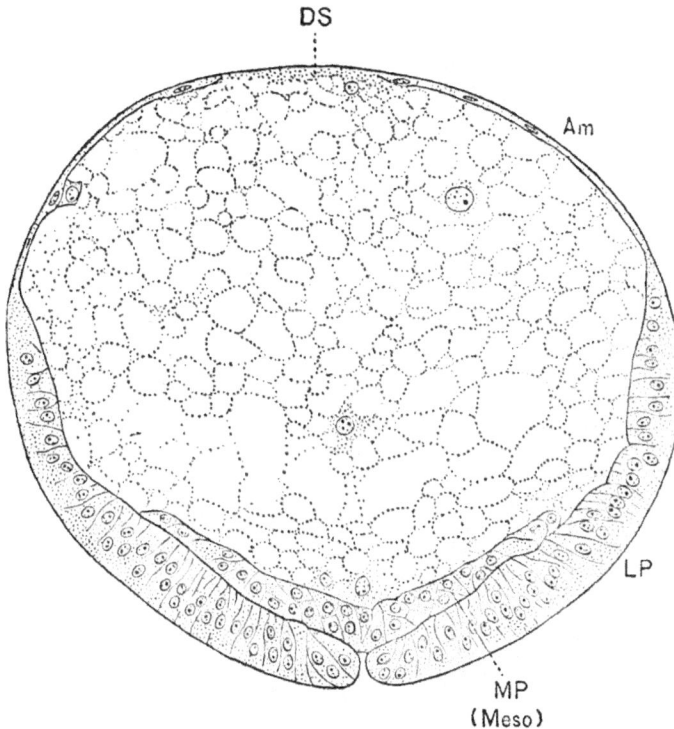

FIG. 21. Transverse section through egg. Stage V-VI, near anterior end, showing the middle plate (*MP*) virtually covered by the lateral plates (*LP*), x 243.

most of the insects studied. Following Korschelt and Heider's classification (1891-3) the different modes of formation of the mesoderm in pterygote insects may be distinguished as follows:

1. The formation of a tube by invagination of a median strip of the ventral blastoderm. This tube, constituting the mesoderm, may be thick-walled and round in section and consequently possessing only a narrow lumen, as in many Coleoptera, or thin-walled and compressed in a dorso-ventral direction, as in *Chalicodoma* (Carrière 1890). This type is probably the commonest and best known of the three. It is found in representatives of widely separated orders, for example: *Hydrophilus* (Kowalev-

ski 1871) *Musca* (Kowalevsky 1886), *Pyrrhocoris* (Graber 1888a), *Chalicodoma* (Carrière 1890).

2. The middle plate remains nearly flat and is overgrown by the edges of the lateral plates, which become free. This type is relatively uncommon but is also found in representatives of widely separated orders, as in *Apis, Sphinx* (Kowalevski 1871), *Pieris* (Bobretsky 1878), *Gasteroidea* (*Gasterophysa*) (Lecaillon, 1897),[9] and, in a slightly modified form in *Forficula* (Heymons 1895).

3. The mesoderm arises by proliferation and immigration from a median area of the ventral blastoderm. A median groove is often also present. This type approaches the conditions obtaining in the Apterygota (Heymons 1897, Uzel 1898) and appears to be especially characteristic of the Orthoptera, being found, for example in *Gryllotalpa* (Korotneff 1885), *Phyllodromia* (Cholodkowsky 1891b), *Gryllus, Periplaneta* (Heymons 1895).

It is accordingly evident that *Apis,* in the manner in which it forms the mesoderm, while agreeing with representatives of remotely allied groups, differs from its near relative *Chalicodoma.* Very little systematic importance can therefore be accorded to these differences. Carrière and Bürger, however, took the opposite view and on the basis of their observations in *Chalicodoma,* declared their belief that the observations of Kowalevski and Grassi in regard to the formation of the mesoderm were incorrect on account of imperfect technique, and that a reinvestigation of *Apis* with modern methods would show that the mesoderm was formed in the same manner as in *Chalicodoma.* It is now sufficiently evident that this assumption was unfounded, at least so far as the eggs destined to form workers or queens are

[8] Carrière (1890, 1897) regards as the middle plate only the inner layer of the two layers of cells which form the mesoderm, the outer layer being considered as formed by the infolding of the mesial portions of the lateral plates. Comparison with the corresponding stages of other insects in which the mesoderm is found by a median invagination, as *Hydrophilus* for example, indicates that this distinction is hardly justified, and that in all cases the term middle plate may properly be considered as including all of the infolded median portion of the ventral plate, and therefore all of the material for the mesoderm.

[9] According to Hirschler (1909a) the mesoderm in *Gasteroidea* is formed by infolding, as in *Hydrophilus*.

concerned. Petrunkewitsch (1902) has published a figure illustrating a transverse section through a drone egg during the
stage of mesoderm formation in which is shown a distinct rolling up of the edges of the middle plate while still attached to
the lateral plates. This condition tends to approach the type
found in *Chalicodoma,* and if it normally exists in the drone
could be readily interpreted as intermediate, connecting types
1 and 2, as illustrated by *Chalicodoma* and *Apis* respectively.

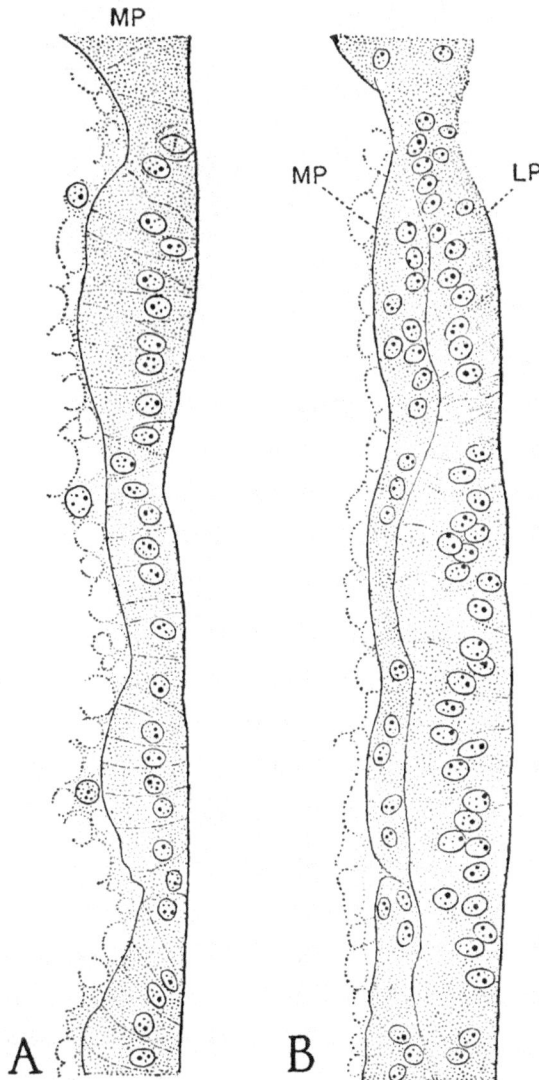

FIG. 22. Longitudinal (sagittal) sections through the anterior region
of the germ band, Stage V, showing segmentation. A is a section along
the mid-line, and includes only the middle plate; B is lateral of the
mid-line. The mesoderm is here now divided into two layers, x 387.

The writer has so far not studied the drone egg, and can therefore not make any statements in regard to it; a rolling up of the edges of the middle plate was however not observed in any of the preparations of the worker eggs.

Coincident with the process of mesoderm formation is another phenomenon, that of *segmentation* or *metamerism*. This is observable as early as Stage V, and on surface view is evident in the form of alternating transverse light and dark bands or zones extending across the ventral plate and appearing first in its anterior region just behind the anterior mesenteron rudiment.

Three of these bands are to be seen in figure V. They make their appearance in rapid succession from in front backwards until at Stage VII they have reached the posterior region of the egg, when at least fourteen of them can be counted. A glance at the next stage, VIII, when the rudiments of the appendages and stigmata make their appearance, suffices to show that the darker zones undoubtedly correspond to the definitive segments of the embryo. Longitudinal sections of Stage VI show that the alternating darker and lighter zones are only the optical expression of alternating thinner and thicker zones in both the middle and lateral plates; in other words the segmentation affects both the future mesoderm and ectoderm (Figs. 22A and B, and 23A and B).

The segmentation of the middle plate is at best rather ill defined, and most marked in the mid-line of those portions about to be covered by the lateral plates (Fig. 22A). The latter show a segmentation corresponding to that of the middle plate and expressed in the same manner, namely by a wavy contour of their inner boundaries when seen in longitudinal section (Fig. 22B). In figure V, where three dark zones are faintly visible at the anterior end of the middle plate, the free edges of the lateral plates at their anterior ends are lobed or scalloped in such a way as to suggest segmentation, the lobes of the two opposite sides corresponding to one another and also to the dark zones of the middle part. The same is true of the lateral plates near their posterior ends at the following Stage, VI. Segmentation (metamerism) thus appears simultaneously in the lateral plates before their union, in the flask stage, and is accompanied by a corresponding segmentation of the middle plate. It begins first

MP   LP   MP   LP

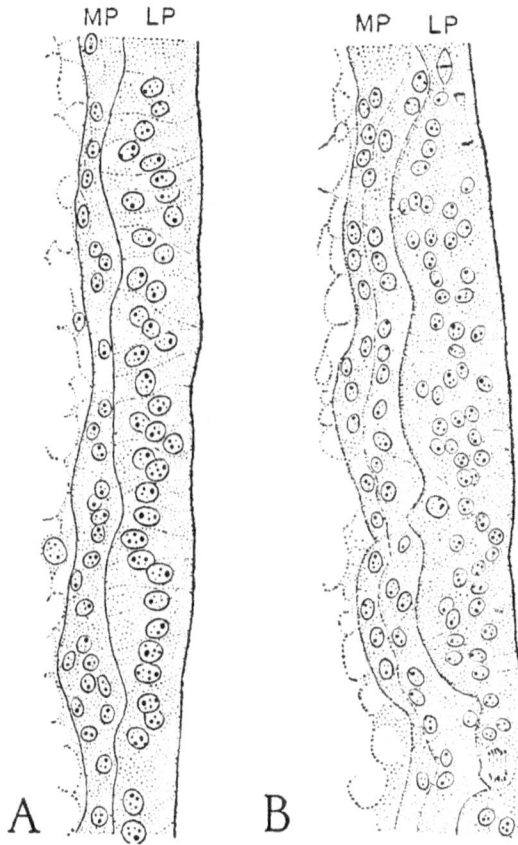

A   B

Fig. 23. Longitudinal (saggittal) sections, through the anterior region of the germ band, Stage VI, showing segmentation. A is a section along the mid-line, B is laterad of the mid-line. The mesoderm is here now divided into two layers, x 387.

at the anterior end of the germ band and progresses backward, following the rule for arthropods in general. Figure 22A illustrates a median sagittal section of Stage V through that portion of the middle plate lying just behind the anterior mesenteron rudiment. Four segments are very plainly visible, much more so than is usual. Figure 22B is taken from the same series and passes laterad of the median plane through one of the lateral plates. It shows very distinctly the segmentation of the future ectoderm, four segments being also represented. The segments are not however always sharply marked off from one another in either the lateral or median plates and are frequently somewhat irregular.

After the middle plate becomes overlaid by ectoderm its lateral regions display a sort of secondary segmentation induced by that of the ectoderm. The mesoderm is here of nearly uniform thickness, in the later stage (VI) double-layered (Fig. 23B), and thrown into a series of low transverse folds by the segmental swellings of the overlying ectoderm. Near the mid-ventral line the segmental swellings of the mesoderm are intact although inconspicuous, but instead of corresponding with those of the ectoderm they fit into the intersegmental depressions.

In general it may be said that up to Stage VIII the segments are not uniformly well defined, and their boundaries not sharp, especially at the two ends of the germ band, so that certain identification of individual segments is difficult if not impossible. Those segments chosen for illustration were unusually well defined. All that can be safely affirmed is the presence of segmentation at this stage.

Bütschli (1870) observed and correctly interpreted these early evidences of segmentation, although neither Kowalevski nor Grassi appear to have noticed them. Bütschli's account is of sufficient interest to quote, and is as follows: "In addition to these primary rudiments of the germinal ridges there are also found the first indication of the segments. I had long overlooked this precocious process, until I investigated the finer structure of the germ band with a high magnification. In contrast to earlier stages this shows no longer the regular cellular structure, but transverse bands, in which the cells are pressed closely together, alternating with the others which are elongated and with their long axis placed in a transverse plane. On closer examination one notices that this condition on the surface harmonizes with the image seen in optical section, which shows swellings alternately with thinner portions, where it is one layered, as well as where it is many layered. Frequently in the contracted portion between two swellings there appears a dark transverse line, which seems to indicate a cleft through the entire thickness of the germ band. The bands described, composed of closely packed cells, form the median portions of the segments, the somewhat larger, more elongated cells lie in the boundary between two neighboring segments. By raising and lowering the microscope tube, I have often persuaded myself that the first bands lie at a higher level than the part lying between them so that ac-

cordingly the external surface of the germ band must also display evidences of segmentation in the form of a faintly wavy
contour" (p. 530).

It is not clear whether the difference just mentioned between
the superficial aspect of the intersegmental and intrasegmental
cells of the germ band applies to the ventral or lateral plates.
Both were examined in the most favorable preparations, but no
differences could be noted. In longitudinal sections of the ventral plate, however, such as that represented by figure 22A, a
considerable difference between the form of intersegmental and
intrasegmental cells is noticeable, the latter being much narrower at their outer ends than the former.

Both Kowalevski (1871) and Heider (1889) described a
precocious segmentation in the egg of *Hydrophilus,* the latter
investigator finding it expressed in the form of transverse folds
which appear even before the middle plate is found. Both
Kowalveski and Heider interpreted these as corresponding with
the future definitive segments. Carrière (1890) found that
the egg of the mason bee (*Chalicodoma*) also showed a precocious
segmentation essentially the same as that just described for the
honey bee. At the "flask" stage, or even earlier, segmentation
makes its appearance on the ventral plate, in surface views of
the egg, as dark transverse bands, alternating with lighter ones,
accompanied by a corresponding lobing of the lateral folds. The
anterior segments appear first, afterward those lying caudad, in
rapid succession. According to Bürger's statement (1897) segmentation appears first in the lateral plates, and only later in the
middle plate. The individual segments become very sharply
marked out, much more so than in the honey bee. Since the
first rudiments of the embryo (antennae, mouth parts, etc.)
appear very early in *Chalicodoma*—as early as the "flask" stage,—
and therefore long before the completion of the union of the
lateral folds, the early identification of the individual segments
is made possible. In both *Chalicodoma* and *Apis* it is important
to note that the segments appearing thus precociously are the
definitive segments of the insect (the so-called microsomites)
and that there is no previous separation of the germ band into
the larger divisions seen first by Ayers (1884) in *Oecanthus,*
afterwards by Graber (1888) in the representatives of several

orders, and termed by him "macrosomites," on the occurrence of which Graber has attempted to construct an elaborate theory of segmentation applicable to the entire arthropod phylum.

2. *Formation of the rudiments of the mid-intestine.*[10]

The mesenteron or mid-intestine in the honey bee is derived from two rudiments, arising at the anterior and posterior ends of the germ band respectively. The *anterior mesenteron rudiment,* owing to its position, is much more readily observed and studied than the posterior mesenteron rudiment, and therefore will be described first.

It will be remembered that the lateral folds end abruptly toward the cephalic pole, leaving vacant an area of the ventral plate about as long as the egg's diameter at this point, corresponding to *the anterior field* of *Chalicodoma* (Carrière and Bürger 1897). At a stage a trifle older than Stage IV, on surface view there may be observed a darker area along the mid-line in this field. This darker area, narrow and rather vaguely outlined at first, rapidly increases in size, density and definiteness until at Stage VI it presents the appearance shown in the figure (VI, *AMR*). *Cf.* also Figs. 24A and B). Its outline is that of a short ellipse, with its longer axis directed lengthwise of the egg; in stained preparations it is deeply stained and very conspicuous. Its width approximates that of the anterior end of the middle plate, with which, at this time, it has come into contact. This is the anterior mesenteron rudiment. Figures 25A to D represent transverse sections through the middle of this rudiment at four successive stages of its development. The first of the series, A, is taken from Stage IV, just before the anterior mesenteron rudiment becomes visible from the exterior, and at its earliest recognizable stage. In the mid-line, over an area whose breadth is about one-eighth of the circumference of the egg at this point the blastoderm cells, hitherto long

---

[10] These have been very widely identified by embryologists as entoderm, and accordingly termed such. Since however there is some doubt as to the correctness of the homology of these rudiments with the entoderm in other classes of animals it has seemed preferable to avoid the use of the term "entoderm" in connection with the development of pterygote insects and, for the present at least, to simply use the term "mesenteron rudiments." See discussion at end of this section.

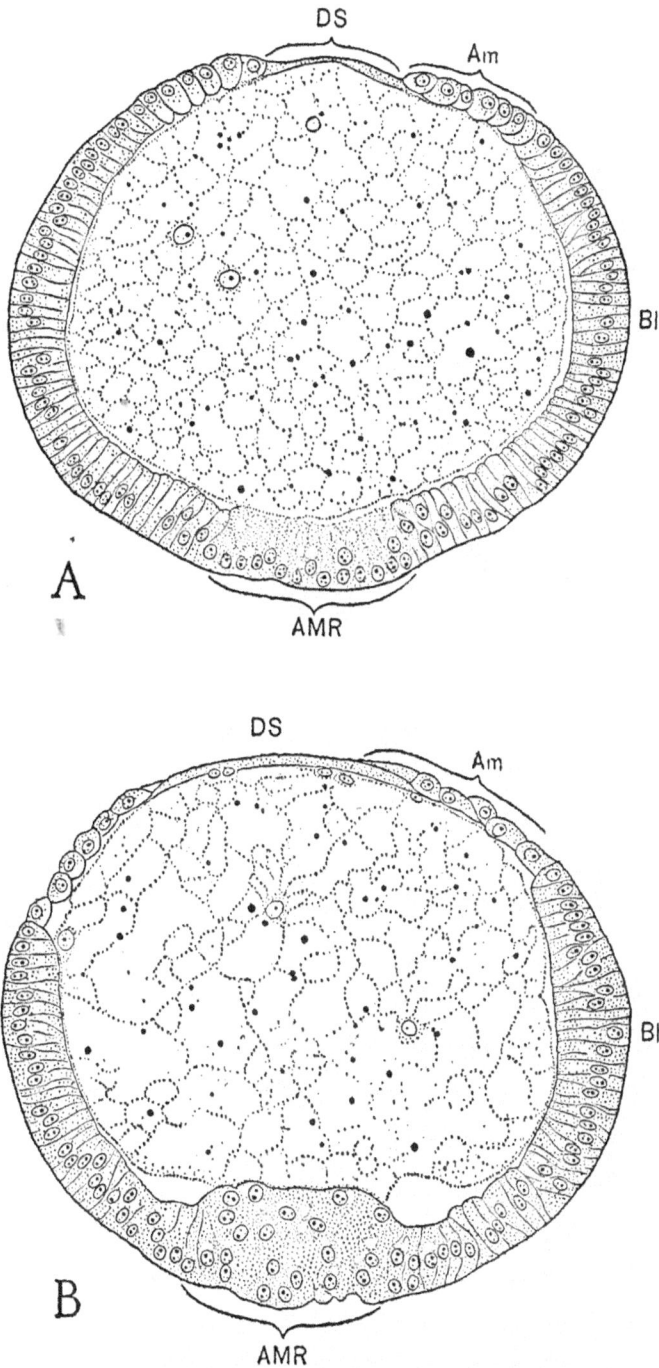

FIG. 24. Transverse sections through the anterior ends of two eggs of Stage IV, intersecting the anterior mesenteron rudiment (*AMR*), illustrating the movement of the lateral blastoderm toward the mid-line during the growth of this rudiment. The development of the amnion in this region is also shown. In A the anterior mesenteron rudiment is barely visible, in B, a slightly older stage, it forms an evident swelling, x 243.

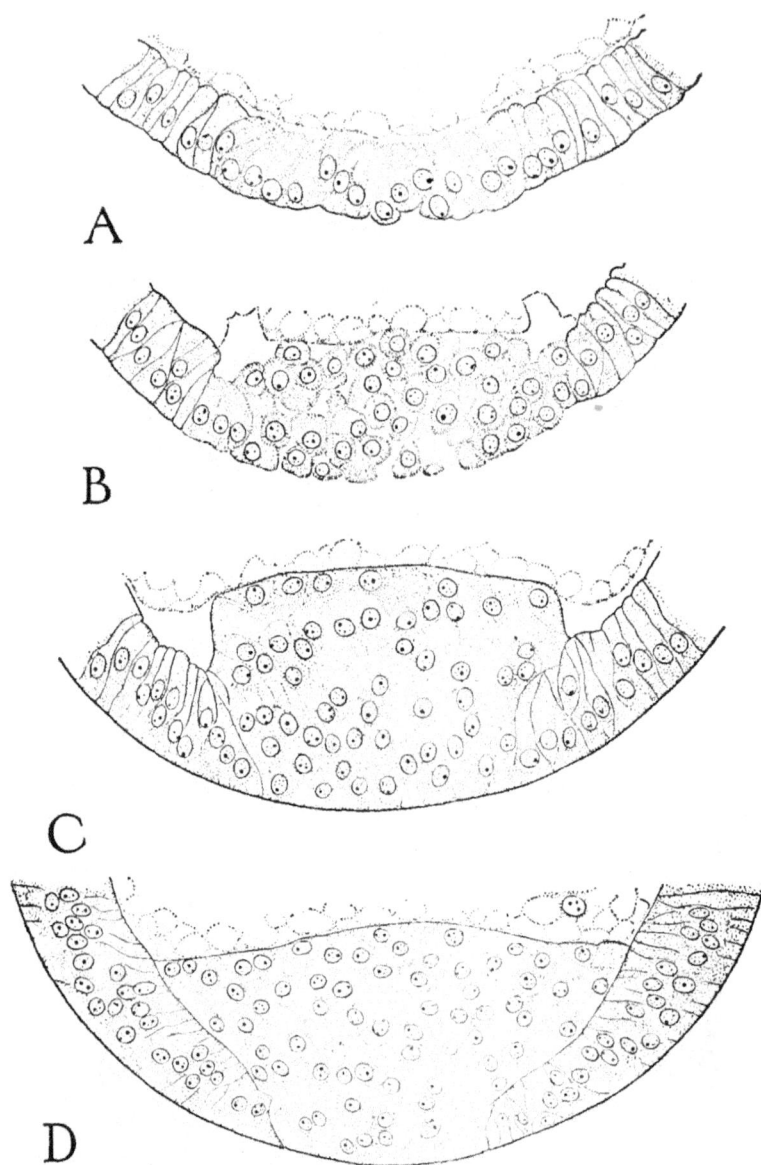

FIG. 25. Transverse sections through the anterior mesenteron rudiments of four eggs, showing four stages in the development of the rudiment. A is from an egg of Stage IV, B and C are from eggs of Stage IV-V, C is from an egg of Stage V-VI, x 290.

prismatic in form, are now seen to be broader, more irregular in form, and loosely arranged. Comparison of sections of a series through the anterior end of an egg at this stage shows that the area involved in the production of the anterior mesenteron rudiment has the form of a narrow triangle lying in the mid-line;

its apex is near the anterior limits of the ventral plate, while
its base joins the anterior end of the middle plate.

In the next figure, 25B, taken from an egg intermediate between Stages IV and V the mesenteron cells are seen to have
increased greatly in number, now forming collectively a low
rounded swelling on the inner side of the blastoderm in the
mid-line. It is at this stage that the anterior mesenteron rudiment first becomes visible from the exterior. The arrangement
of the component cells is still loose, numerous interstices remaining between them. In form the cells are, generally speaking, ovoid or rounded, although somewhat irregular. On one
side of the nucleus a vacuole is frequently to be seen, as in the
adjacent blastoderm cells, which they closely resemble in all
respects except that of form. As the cells of the anterior
mesenteron rudiment continue to increase rapidly in number
(Fig. 25C), the rudiment increases correspondingly in thickness
and breadth, so that it spreads out laterally. At the same time,
its cells become compactly arranged and polyhedral in form,
due to mutual pressure doubtless caused by the necessary displacement of the underlying yolk. Figure 26A represents a
longitudinal section through the anterior mesenteron rudiment at
about this period (Stage IV-V). The rudiment extends cephalad to within three or four cells of the anterior end of the
ventral plate, where it ends abruptly, its cephalic face perpen-
dicular to the surface of the blastoderm, leaving in front of the
rudiment a cavity bounded on its inner side by the yolk, on its
outer by the amnion (*Am*) and by the short stretch of unmodified blastoderm cephalad of the mesenteron rudiment. Caudad
the latter diminishes gradually in thickness to join the middle
plate (*MP*).

Figure 25D shows the rudiment after the lateral folds have
commenced to unite (Stage V-VI) being drawn from the same
preparation as figure 21. The anterior mesenteron rudiment
has now increased considerably in the number of its cells and in
its lateral extent, coming into close contact with the inner surface of the lateral blastoderm on both sides of the mid-line. In
addition to its increase in size another change is becoming evident, which has the appearance of an encroachment of the
lateral unmodified blastoderm (ectoderm) on the external layers

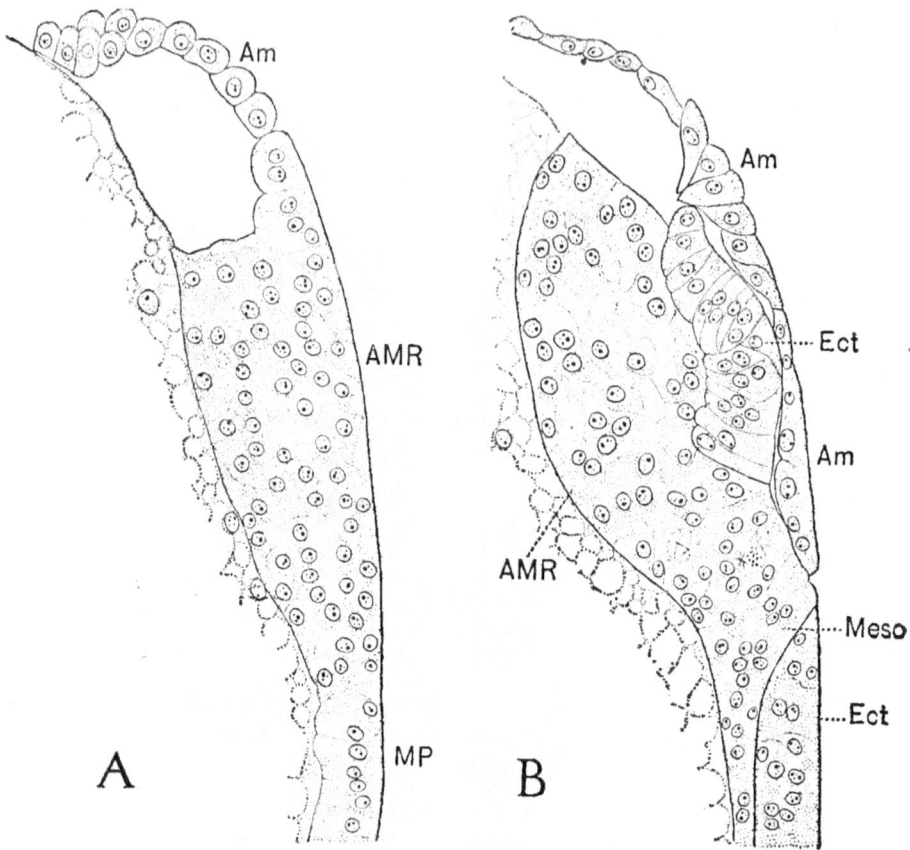

FIG. 26.   Median sagittal sections through the anterior mesenteron rudiments of two eggs, illustrating the development of the rudiment and its relations to the adjacent ectoderm (*Ect*) and mesoderm (*Meso*). The formation of the amnion (*Am*) in this region is also shown. A is from an egg of Stage IV-V, B from an egg of Stage VI, x 290.

of the anterior mesenteron rudiment, thereby gradually restricting its superficial area, and leading finally to the complete covering of its external surface by ectoderm, with the exception of a circular area near its posterior border (Fig. 26B). The precise nature of this process is uncertain. Carrière (1890, 1897) has stated that in the mason bee it consists in a centripetal differentiation of the superficial cells of the rudiment into prismatic ectoderm cells, but in the honey bee there are certain concomitant phenomena not evident in the mason bee, which make it seem possible that the covering of the anterior mesenteron rudiment is brought about by a simultaneous mesiad movement of the two halves of the ventral plate separated by the rudiment. This covering by the ectoderm is completed first at the anterior

narrow end of the superficial part of the rudiment; it then
progresses rapidly both caudad and mesiad at the same time.
This process occupies the interval between Stages V and VII.
It is important to note that the mesiad progress of the lateral
ectoderm over the anterior mesenteron rudiment and the approximation of the anterior ends of the lateral plates over the
middle plate takes place at the same time and goes on at about
the same rate.

The nuclei of the mesenteron cells, spherical in form from the
inception of the rudiment, now begins to be more clearly distinguishable from those of the surrounding tissues, being distinguished not only by their circular outline but also by their
somewhat greater size and paleness. These differences serve as
useful means of identification during the succeeding stages of
embryonic development.

The form and relations of the anterior mesenteron rudiment to
the adjacent parts at Stage VI when the ectoderm is completely
formed over the anterior half of the germ band is shown in
figure 26B, which represents a median longitudinal section
through the anterior end of the germ band. The anterior mesenteron rudiment is here seen to be thick lenticular in form, its
anterior edge projecting out from beneath the anterior margin
of the overlying ectoderm. Betweeen the latter and the ectoderm (*Ect*) which overlies the mesoderm (*Meso*) is a rather
wide gap representing a rounded break in the continuity of the
ectodermal covering of the germ band. This marks the place of
origin of the future mouth. Through this gap or orifice, cells of
the posterior edge of the anterior mesenteron rudiment are seen
to come to the external surface, filling the space with a plug of
cells. A similar relation between the anterior mesenteron rudiment and the overlying ectoderm occurs in *Chalicodoma* (Carrière and Bürger 1897, and between both mesenteron rudiments
and ectoderm in *Gasteroidea* (Hirschler 1909a). This plug
however is not composed exclusively of the cells of the anterior
mesenteron rudiment, since a few mesoderm cells, distinguished
by their smaller nuclei, are seen in its posterior portions. Posteriorly the anterior mesenteron rudiment is very closely united
with the anterior end of the mesoderm, no sharp line of separa-

tion being visible, the only distinguishing character being the differing size of the nuclei.

The source and manner of origin of the anterior mesenteron rudiment remain to be considered. The possibility of its derivation from yolk cells is excluded by the simple fact that yolk cells are never present in this region in sufficient numbers to form such a structure, moreover, the close relation of the rudiment with the blastoderm and the similarity of its cells to the cells of the blastoderm make the assumption of any other source than the blastoderm impossible.

Carrière (1890) and Carrière and Bürger (1897) state that the mesenteron rudiments in the mason bee arise from the blastoderm by proliferation. In the honey bee however during the earlier stages, comprising the period of its most rapid growth, when the greater portion of its mass is formed, cell divisions are virtually absent in that part of the ventral plate from which the mesenteron rudiment arises. This is particularly significant, when it is considered that this period is very brief, extending from Stages IV to V. Since, therefore, the anterior mesoderm rudiment does not arise by proliferation it must be assumed that it arises by immigration. This is strongly suggested by the appearance seen in sections like that represented in figure 25B. At this stage the cells bordering the anterior mesenteron rudiment laterally are plainly seen to intergrade with those of the blastoderm. This is also true of the section represented in figure 25C, but is perhaps less evident.

Two phenomena coincident with the formation of the anterior mesenteron rudiment take on a special significance when considered in association with the facts just mentioned. These are: first, that the appearance of the anterior mesenteron rudiment and the depression of the middle plate are contemporaneous; second, that the two lateral halves of the ventral plate bounding the mesenteron rudiment appear to move mesiad during the period when the lateral plates are coming together to cover the middle plate, or, to express it in a different way, the lateral halves of the ventral plate bounding the anterior mesenteron rudiment behave toward the latter much as the lateral plates behave toward the middle plate. Figures 24A and B illustrate this point. Figure 24A is drawn from a section through the

anterior end of an egg at Stage IV; figure 24B is drawn from
a section through the same region at a stage about half way between Stages IV and V. The mesiad movement of the two
lateral halves of the ventral plate in this region during the intervening period becomes at once evident, since the area covered
by the dorsal strip and the amnion-forming cells has become
greatly extended transversely, accompanied by a flattening of
the amnion-forming cells. The possibility that this movement
is only an apparent one, and caused by a contraction of the
lateral blastoderm is excluded, since the blastoderm cells are
not perceptibly lengthened in the section taken at the later
stage, as would be the case if an actual contraction had taken
place.

It appears then, in view of the facts outlined above, that the
anterior mesenteron rudiment is derived from the cells of the
ventral blastoderm by immigration, and that this rudiment is
continuous with, and comparable to, the ventral plate, in so far
as both are formed contemporaneously from the middle section
of the ventral plate, accompanied by a mesiad movement of its
lateral sections. Moreover, the anterior mesenteron rudiment
and the ventral plate are directly continuous with one another,
the chief differences between them being that the former arises
as a heap of cells not sharply marked off from the remainder
of the blastoderm, while the latter arises as a solid flat section of
the blastoderm, discontinuous with the lateral blastoderm and
separated from it by a sharp break.

Cell division, although apparently playing no part in the formation of the anterior mesenteron rudiment during its earlier
stages, steps in during the later stages, when mitotic figures are
frequently seen, being most abundant in the posterior part of the
rudiment. Two mitotic figures are evident in this region in
figure 26B.

The manner in which the ectoderm finally closes over the anterior mesenteron rudiment is, as before mentioned (p. 60), obscure, and repeated efforts finally to decide this question were
without definite result, nevertheless, since it is certain that the
lateral blastoderm bounding the anterior mesenteron rudiment
moves mesiad during the early stages of the rudiment it seems
possible that the final covering of the external surface is brought

about in this manner. Sections like that represented in figure
25D lend color to this supposition, since here the mesial edges
of the lateral blastoderm are marked off from the mesenteron
cells with a considerable degree of sharpness.

The *posterior mesenteron rudiment* is much more difficult to
study than the anterior rudiment, for two reasons; first, being
formed near the posterior pole, in the course of its development
the lengthening of the germ band carries the rudiment over the
rounded posterior end towards the dorsal side, so that sections
normal to the surface of the germ band at this point are seldom
obtained, and second, the posterior end of the egg, being the
one by which it is attached to the floor of the cell of the comb,
is frequently damaged. For the first of these reasons, longitudinal sections are more informing and more useful than transverse
sections, which are commonly cut at right angles to the long
axis of the egg, and therefore only rarely intersect the posterior
mesenteron rudiment at right angles to its own long axis. As a
matter of fact, scarcely any satisfactory transverse sections
through this rudiment were obtained.

The development of the posterior mesenteron rudiment is
fundamentally identical with that of the anterior mesenteron rudiment, but differs greatly in details. Unlike the anterior mesenteron rudiment, the posterior mesenteron rudiment is ordinarily not
evident on preparations of entire eggs, except during the concluding stages of its development, when it appears as a deeply stained
discoid mass at or dorsal to the caudal pole. Prior to Stage V
the ventral plate extends over the caudal pole to the dorsal surface as a single layer of low cells, rather irregular in form, and
rounded on their external surface. This layer is frequently in-
terrupted by short gaps which leave bare the yolk beneath. At
Stage V the first indication of the posterior mesenteron rudiment
becomes visible as a slightly thickened area of the blastoderm
just cephalad of the caudal pole, on the ventral surface (Fig.
27A, *PMR*). The cells composing this thickening are now long
prismatic in form, instead of low and rounded as before. The
posterior portion of the ventral plate has apparently meanwhile
contracted, at least in a longitudinal direction, since it now extends only a short distance dorsad to the caudal pole. At the
stage following, Stage VI (Fig. 27B), the posterior mesenteron

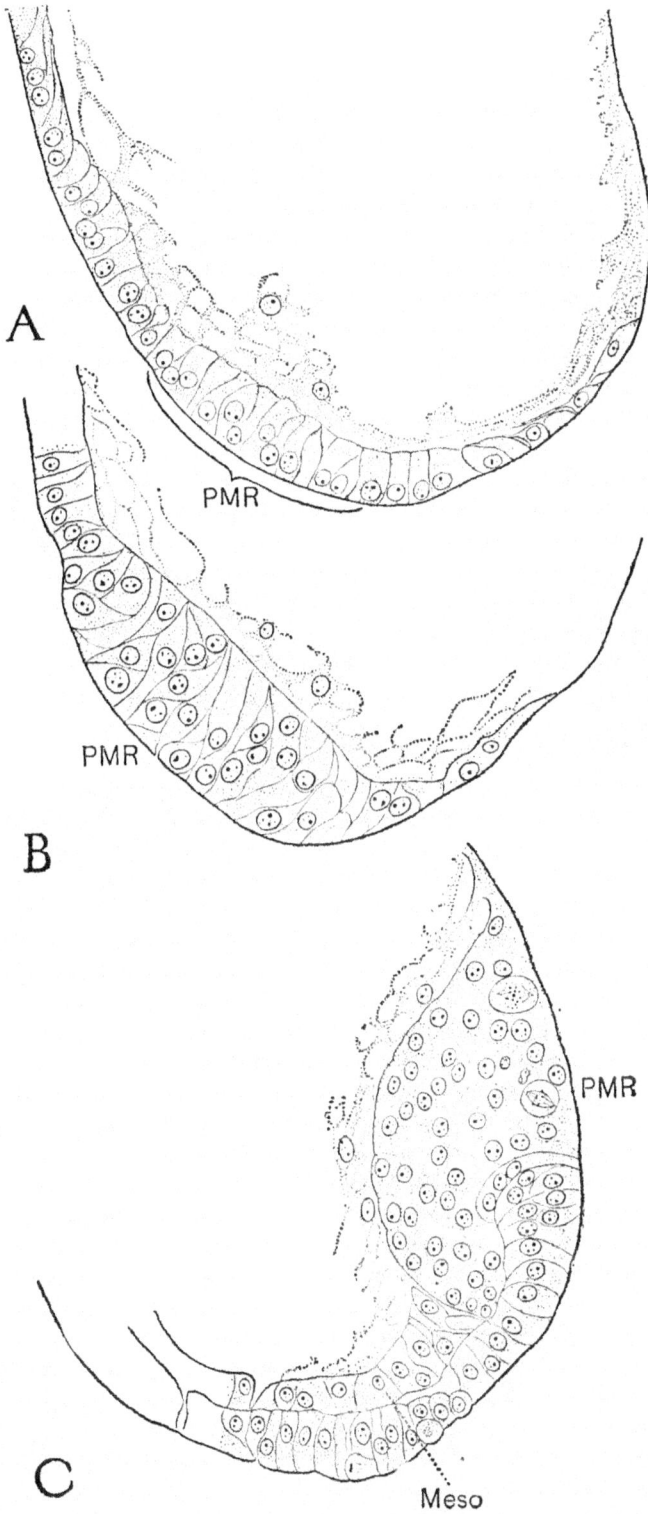

FIG. 27. Median sagittal sections through the posterior ends of three eggs, to illustrate the formation of the posterior mesenteron rudiments (*PMR*). A is from an egg of Stage V, B from an egg of Stage VI, C from an egg of Stage VII, x 290.

rudiment is seen to have increased greatly in thickness. It now forms a discoid mass lying on the ventral surface of the egg, its posterior boundary lying almost precisely at the caudal pole. Its extent is approximately the same as that of the anterior mesenteron rudiment. As the illustration shows the posterior mesenteron rudiment is now composed of long irregularly bent and curved fusiform cells, some of which at least extend throughout the entire thickness of the layer. At the next stage, Stage VII (Fig. 27C), four changes are seen to have taken place. (1) The rudiment, by a lengthening of the germ band, has been shifted around the caudal pole of the egg to the dorsal surface. (2) It is no longer composed of long fusiform cells, but of polyhedral cells precisely like those of the anterior mesenteron rudiment. (3) The ectoderm now covers its anterior half. (4) It has increased greatly in thickness, which approximates one-half its diameter. The question as to the manner in which the posterior mesenteron rudiment becomes covered by ectoderm is even more difficult to answer than in the case of its counterpart at the anterior end. That this covering is brought about by extension of the ectoderm at the expense of the superficial cells of the rudiment is strongly suggested by the section represented in figure 27C. It is certain at least that the extension takes place in an antero-posterior direction with regard to the embryo itself,—and that the ectodermal covering of the rudiment is absolutely continuous with that of the lateral plates, since in one series (that from which Fig. 28 was drawn) the cleft still separating the lateral plates can be followed around the caudal end of the egg to the posterior limit of the ectoderm. Figure 28 represents a transverse section through the posterior end of an embryo of Stage VII, and intersects the posterior mesenteron rudiment about midway of its length. On the ventral side are seen the lateral plates (*LP*) still separated by a narrow cleft, and lying above them, the middle plate (*MP*), or mesoderm. The dorsal half of the section is occupied by the massive posterior mesenteron rudiment (*PMR*), connected on each side with the lateral plates by a thin sheet of cells, the amnion. This figure illustrates an important difference between the anterior and posterior mesenteron rudiments. While the former is produced only by a relatively narrow median strip of blastoderm, the lat-

PMR

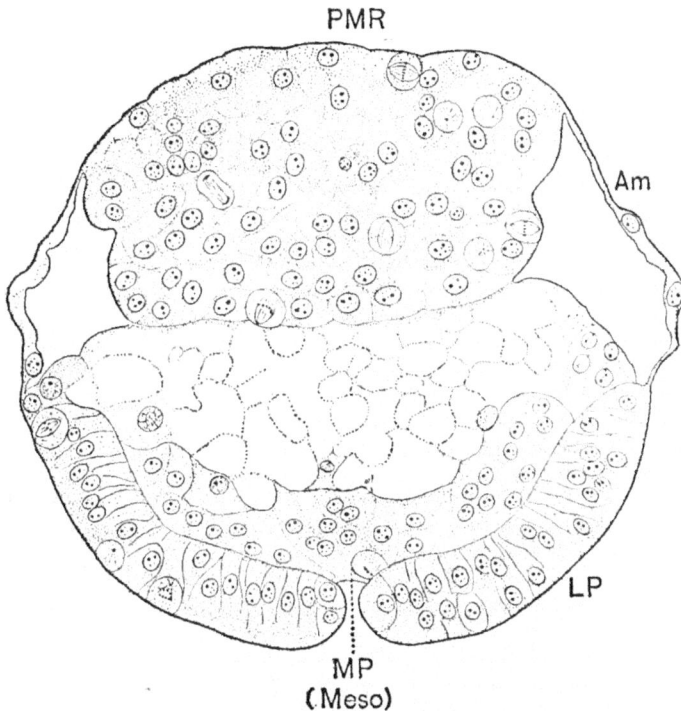

Am

LP

MP
(Meso)

FIG. 28. Transverse section through the posterior end of an egg of Stage VII, showing the posterior mesenteron rudiment (*PMR*), and also the relations of this rudiment and of the lateral plates (*LP*), middle plate (*MP*), and amnion (*Am*), to each other, x 290.

ter demands for its production the entire posterior end of the ventral plate, leaving at the sides no undifferentiated blastoderm (ectoderm). In the case of the anterior mesenteron rudiment the ectodermal covering progresses mesiad from the mesial edges of the two halves of the ventral plate bounding the rudiment laterally. In the case of the posterior mesenteron rudiment the ectodermal covering progresses caudad from the posterior limits of the lateral plates, appearing as if they were extended backward to cover the posterior mesenteron rudiment. The external surface of the latter is, however, never entirely covered, its posterior end (relative to the embryo) apparently always remaining uncovered. Moreover the ectodermal covering forms a continuous and unbroken sheet to its caudal limits. There is no break in its continuity corresponding to the future proctodaeal invagination and comparable to the uncovered area of the anterior rudiment. This difference is of importance, since it foreshadows the differing relations of the fore- and the hind-intestine to the mid-intestine in the larva, when communication between the mesenteron (mid-intestine) and hind-intestine is completely cut off.

The source of the posterior mesenteron rudiment is obviously the posterior end of the ventral plate; its mode of origin is much more clear than in the case of its counterpart at the anterior end. In the earlier stages of the formation of the posterior mesenteron rudiment, as in the anterior rudiment, mitotic figures are rare, so that cell division cannot be regarded as an important factor in the earlier stages of its formation. The posterior end of the ventral plate shares in the general decrease in breadth associated with the formation of the germ layers, moreover, it also contracts in a longitudinal direction. Decrease in the length and breadth of the ventral plate at its posterior end together with a corresponding elongation of its component cells may then be safely set down as the factors first concerned in the formation of the posterior mesenteron rudiment. During the later stages (VI-VII) cell division becomes an increasingly important factor in its growth, as in the anterior rudiment. At Stage VII mitotic figures are exceedingly abundant, as figures 27C and 28 show.

In conclusion then it may be said that both the anterior and the posterior mesenteron rudiments are formed from the blastoderm of the ventral plate by a movement inward of its cells. In the case of the anterior rudiment the migration takes place by the detachment of cells from a limited area in the mid-line. The rudiment is afterward covered over either by a movement of the lateral halves of the remaining blastoderm toward the midline, or by additions to their mesial borders from the superficial cells of the rudiment itself. In the case of the posterior rudiment the entire posterior end of the ventral plate is involved, being moulded by changes, first in the form and next in the arrangement of its cells, to form the posterior mesenteron rudiment. The ectoderm, which later covers it, is continuous with the posterior ends of the lateral plates, and appears as a caudad extension of them, although it cannot be stated with certainty whether they are formed by an actual caudad extension or by the addition of new material from the external surface of the rudiment.

It cannot fail to be noticed that there is a fundamental similarity between the formation of the mesenteron rudiments and of the mesoderm, since all three arise by an inward movement of cells of the ventral plate. Moreover, the latter is continuous cephalad and caudad with the mesenteron rudiments during the period of their formation.

The origin of the mesenteron in the honey bee was first discussed by Bütschli (1870). Bütschli believed it to be produced like the blastoderm by the so-called "free cell formation."[11] Referring to a stage nearly corresponding to that numbered VIII in this paper he says (pp. 540-541) "the formation of the mid-intestine has already commenced, yet, as I often observed in much more advanced embryos, only on the dorsal side of the yolk; seen in optical section, it is a single layer of close pressed, yellowish cells. It appears to me that this cell layer takes its origin near the ends of the germ band, since its thickness is here greater than in the middle region. I have not succeeded in observing its beginning, but in regard to it must conclude with Zaddach and Weismann that the mid-intestinal wall develops by free cell formation and not by delamination of an inner cell layer." It is interesting however to find that Bütschli both noted and figured the anterior mesenteron rudiment, although without recognizing it as such. He says concerning it (p. 529): "In connection with the regression of the germ band at the anterior pole is a thickening found there, concerning the true significance of which I am not sure; seen *en face* it appears rounded, in profile, it projects inward, yet I believe that the hemispherical swelling represented in figure 10B is not in reality such, but that the lateral extension of the thickening has produced an optical illusion." The figure referred to corresponds to a stage about midway between Stages V and VI.

Kowalevski (1871) believed that the mesenteron in the bee owed its origin to the inner or splanchnic layer of the mesoderm. He was evidently led to this belief by the fact that in sections the

---

[11] This consists in the spontaneous formation of nuclei within a protoplasmic matrix. It is hardly necessary to say that such a process does not exist, every cell arising by division from a preexisting cell. The belief was due to the imperfect technique employed by the investigators.

mesenteron at later stages is seen lying close beneath the lateral portions of the mesoderm.

Grassi (1884) gave an essentially correct account of the origin of the mesenteron in the honey bee, and one which has been generally accepted as such. His observations of the development of the anterior mesenteron rudiment were much more satisfactory and complete than those of the posterior rudiment. The substance of his statement in regard to it is as follows: After the formation of the mesoderm (middle-plate) has begun, the median part of that area of the blastoderm anterior to the furrows (lateral folds) becomes many layered, with the exception of its anterior margin. Later, beginning at the lateral margin of the many-layered portion, and perhaps also at its anterior margin, the superficial layer is separated from the deeper layers. This superficial layer is continuous with the one-layered blastoderm at the anterior pole, and is also continuous caudad with the ectoderm and is itself ectoderm. The deeper layers are continuous caudad with the mesoderm and are themselves mesoderm. The formation of the posterior mesenteron rudiment was supposed to be similar to that of the anterior rudiment. One entire plate (Pl. VI) containing thirty-six figures of sections is devoted to the development of the mesenteron rudiments. It is to be noted, however, that Grassi makes no statement as to the manner in which these rudiments are produced, and also failed to note the break in the ectoderm covering the anterior mesenteron rudiment, where the mouth is formed later. Moreover, his belief that these rudiments are to be interpreted as mesoderm is, in view of subsequent investigation, scarcely justified; nevertheless his account of the origin of the mesenteron rudiment is probably the most important part of Grassi's paper, and also the one which attracted most attention.

Carrière and Bürger's (1897) account of the formation of the mesenteron rudiments of the mason bee is substantially the same as that of Grassi, differing from it principally in interpretation. According to these investigators the anterior and posterior mesenteron rudiments arise from proliferating areas of the undifferentiated blastoderm at the two ends of the ventral plate, respectively cephalad and caudad to the middle plate and independent of it. These rudiments constitute two large hemispherical cell masses of which the posterior is the larger, their convex surfaces directed inward. Later the superficial layer of these rudiments is modified to form ectoderm, with the exception of a small area situated over the center of the anterior rudiment, which remains unmodified and actively proliferating, and which is to form the floor of the stomodaeal invagination. This area corresponds to a similar area over the anterior mesenteron rudiment of the honey bee, described above. Carrière and Bürger, in opposition to Grassi, insist strongly on the independence of the mesenteron rudiments from the mesoderm, and consider those parts of the ventral plate from which the mesenteron rudiments take their rise as purely blastodermal.

Only one other paper in the social Hymenoptera remains to be mentioned in this connection, that of Dickel (1904) on the honey bee. This is unique in that it seeks the origin of the "entoderm" in a peculiar discoid cell mass appearing at the anterior end of the egg during the earlier stages in the formation of the germ layers, and derived from yolk cells, and therefore turned "yolk plug" or "yolk syncitium." A corresponding "yolk plug" was assumed by Dickel to exist at the posterior end of the egg. These cell masses are supposed to be later carried inward by invagination and to constitute the anterior and posterior mesenteron rudiments. In a succeeding section the origin and fate of the "yolk plug" of Dickel will be discussed in detail; it is sufficient to state here that it has no connection with the mesenteron rudiments. The invagination figured by Dickel at the anterior end of the egg can readily be construed as an artifact, since such infoldings are very common in eggs of the bee which have not been properly handled and are frequently produced by the osmotic pressure of a clearing agent, such as cedar oil, when incautiously used.

The origin of the mesenteron of insects has for the past forty years been the subject of numerous investigations and also a prolific source of discussion, from which the partisan spirit has not been altogether absent. In the history of embryological research there is perhaps no problem about which there has been a greater diversity of opinion, and it is a regrettable fact that even at the present day investigators of this subject are still arrayed against one another in opposing camps. This is still more regret-

table since it is a recognized fact that much of this want of agreement rests on differences of interpretation. These various observations and interpretations have been frequently discussed, often
at great length, in the various textbooks and the papers which
have dealt with this phase of insect embryology, so that a prolonged review of the individual papers appears superfluous, but
in order to gain an insight into this perplexing subject another
means has suggested itself, which is embodied in the following
classified list or table of the different investigators who have dealt
with the origin of the mesenteron. It shows (1) the particular
view adopted, (2) its adherents in order of the dates of publication of their papers, (3) The genus or genera of insects on
which the observations were made. It is realized that such a
table fails in many instances to represent the differences as regards details, but on the other hand these could not be brought
out except in a prolonged discussion, and in any event should
be sought in the original paper. It is also realized that this
table may not be altogether wanting in inaccuracies and omis-
sions, but a conscientious effort has been made to reduce these
to a minimum by consulting the original papers wherever possible. If this table does nothing more it will serve at least to
throw into relief the opposing views and interpretations, and
also to illustrate their diversity.

(I) Mesenteron derived from yolk cells.
Dohrn (1866, 1876).
Bütschli (1870) *Apis*.
Mayer (1876).
Bobretzsky (1878) *Pontia* (*Picris*).
Graber (1878).
Balfour (1880).
Hertwig (1881).
Weismann (1882) *Rhodites*.
Tichomiroff (1882) *Bombyx*.
Ayers (1884) *Oecanthus, Telias*.
Patten (1884) *Neophalax*.
Korotneff (1885) *Gryllotalpa*.[12]

[12] Korotneff believed that only the embryonic mesenteron was formed
by yolk cells, the functional or larval mesenteron owing its origin to
blood cells.

Will (1888, 1888a) *Aphis.*

Tichomirowa (1890) *Chrysopa.*

Tichomiroff (1890) *Calandra.*

　　　　(1892) *Bombyx* and *Calandra.*

Tichomirowa (1892) *Pulex.*

Heymons (1897) *Lepisma* and *Campodea.*

Claypole (1898) *Anurida.*

Tschuproff (1903) *Epitheca* and *Calopteryx* (Median section of mesenteron only.)

Dickel (1904) *Apis.*

(11) Mesenteron derived from the lower layer (mesoderm, entomesoderm, primary entoderm).

1. Derived solely from anterior and posterior sections.

Grassi (1884) *Apis.*

Kowalevsky (1886) *Calliphora* (*Musca*).

Graber (1889) *Lucilia, Calliphora* (anterior rudiment only) *Lina.*

Wheeler (1889) *Blatta, Leptinotarsa* (*Doryphora*).

Ritter (1890) *Chironomus.*

Heider (1888) *Hydrophilus.*

Karawaiew (1893) *Pyrrhocoris.*

Kulagin (1897) *Platygaster.*

Escherisch (1900) *Calliphora.*

Petrunkewitsch (1902) *Apis.*

Schwangart (1904) *Endromis, Zygaena.*

Nusbaum and Fulinski (1909) *Gryllotalpa.*

2. Derived from anterior and posterior sections and also from a median section.

Nusbaum (1886) *Periplaneta.*

Nusbaum and Fulinski (1906) *Phyllodromia.*

Hirschler (1905) *Catocala.*

　　　　(1909) *Donacia.*

　　　　(1909a) *Gasteroidea* (*Gasterophysa*).

3. Derived from anterior and posterior sections of mesoderm and also from splanchnic (inner) layer.

Heider (1885) *Hydrophilus.*

Cholodkowsky (1891c) *Phyllodromia* (*Blatta*).

4. Derived from splanchnic layer of mesoderm by delamination of two lateral bands.

Kowalevski (1871) *Apis, Hydrophilus.*

Tichomiroff (1879) *Bombyx.*

Cholodkowsky (1888) *Phyllodromia (Blatta).*

Graber (1888a) *Stenobothrus, Lina.*

5. Derived from splanchnic layer of mesoderm by delamination and also from a median section.

Nusbaum 1888) *Meloë.*

6. Derived from median section of mesoderm only.

Hammerschmidt (1910) *Dixippus.*

(III) Mesenteron derived from proliferations of the blind inner ends of the stomodaeal and proctodaeal invaginations.

Ganin (1874).

Witlaczil (1884) *Drepanosiphum (Aphis).*

Voeltzkow (1888, 1889, 1889a) *Calliphora (Musca), Melolontha.*

Graber (1889) *Lucilia, Calliphora* (posterior rudiment only).

Graber (1891) *Stenobothrus.*[13]

Graber (1891b) *Gryllotalpa, Meloë.*

Korotneff (1894) *Gryllotalpa.*

Heymons (1894, 1895) *Forficula, Gryllus, Gryllotalpa, Periplaneta, Phyllodromia (Blatta), Ectobia.*

Heymons (1897a) *Bacillus.*

Lecaillon (1898) *Clytra, Gasterophysa, Chrysomela, Lina, Agelastica.*

Rabito (1898) *Mantis.*

Schwartze (1899) *Lasiocampa.*

Toyama (1902) *Bombyx.*

Pratt (1900) *Melophagus.*

Deegener (1900) *Hydrophilus.*

Tschuproff (1903) *Epitheca* and *Calopteryx* (anterior and posterior sections of mesenteron only).

Czerski (1904) *Meloë.*

Hirschler (1905) *Catocala.*[14]

Saling (1907) *Tenebrio.*

---

[13] Graber believed that cells were also added to the mesenteron from the splanchnic layer of the mesoderm.

[14] Hirschler believes that in *Catocala* a median section of the lower layer also contributes cells to the mesenteron.

Friederichs (1906) *Donacia.*

(IV) Mesenteron derived, independent of the mesoderm, from two proliferating areas of the blastoderm, one at each end of the germ band, corresponding to the future location of the stomodaeum and proctodaeum, respectively.

Carrière and Bürger (1897) *Chalicodoma, Tenebrio.*

Noack (1901) *Calliphora.*

(V) Mesenteron derived from cells migrating inward from thickenings or islands of the blastoderm.

Uzel (1897, 1898) *Lepisma, Campodea.*

The importance accorded by all investigators of insect embryology to the question of the origin of the mesenteron is due of course to its relation to the germ layer theory. According to this theory, which is based on a large number of observations on the development of various animals, the material from which the mesenteron is formed should correspond to entoderm and the efforts of practically all investigators of this problem have been bent toward establishing this homology, and to derive the conditions found in the insect egg from the typical gastrula. This has proved to be an exceedingly difficult task. Thirty years ago Weismann (1882) wrote (p. 81): "It becomes more and more evident, that nowhere in the entire animal kingdom is the ontogeny so distorted and coenogenetically degenerate, as in the insects, so that scarcely anywhere are the germ layers so difficult to recognize as here." Time has proved the truth of these statements.

Prior to 1884 nearly all of the investigators of insect embryology were divided into two camps; either they followed Dohrn (1866) in deriving the mesenteron from the cells remaining in the yolk, or followed Kowalevski in deriving the mesenteron from the inner wall (splanchnic layer) of the mesodermic somites. Only one exception is to be noted: Ganin (1874) described the mesenteron as derived from the inner ends of the ectodermal proctodaeal and stomodaeal invaginations. In these three views are contained the germs of all the later theories of the origin of the mesenteron.

In 1884 Grassi's paper on the development of the honey bee appeared, in which the bipolar origin of the mesenteron from the lower layer (mesoderm) was first demonstrated. Kowalevsky in 1886 published a brief paper on the origin of the mesenteron rudiments in *Musca*. In this insect the mesenteron rudiments were formed, much as Grassi had described them in *Apis*, from the anterior and posterior ends of the middle plate (mesoderm). Kowalevsky constructed an ingenious theory to account for the origin of the germ layers in insects, comparing the conditions which he found in the insect embryo with those occurring in a marine invertebrate, *Sagitta*. He regarded the insect egg, at the time of the formation of the germ layers, as comparable to a gastrula so much stretched out or elongated that the entoderm (mesenteron rudiments) was pulled into two halves. The views of Grassi (see above p. 70) and Kowalevsky were later accepted by Wheeler (1889), Cholodkowsky (1891), and by Heider in Korschelt and Heider's textbook, and recently in a modified form, by Nusbaum and his pupils (II. 2).

These results did not however find universal acceptance, since Witlaczil, Voeltzkow and Graber, like Ganin, contended that the mesenteron arose from the blind inner ends of the proctodaeum and stomodaeum. In 1895 Heymons, in his handsome monograph on the development of the Orthoptera and Dermaptera, devoted especial attention to the development of the mesenteron, and also found that it was derived from the blind inner ends of the stomodaeal and proctodaeal invaginations. This work has had a wide influence and Heymons' results have been confirmed by a number of investigators in the Coleoptera, Lepidoptera and Orthoptera.

Heymons recognized more fully than his predecessors the theoretical difficulty involved in deriving the mesenteron from ectoderm, since identification of the mesenteron of insects with entoderm is thereby precluded. Heymons boldly met the difficulty by supposing that the functional mesenteron of pterygote insects is of comparatively recent origin, and that the original entoderm is now represented by the yolk cells, which therefore may be considered as constituting a vestigial or degenerate mesenteron. This view received support by Heymons' discovery (1897) that in *Lepisma*, a primitive apterygote insect, the functional mesenteron is actually formed from yolk cells and by Madame Tschuproff-Heymons' discovery (1903) that in the Odonata the mesenteron is formed in part by yolk cells and in part by ectoderm derived from the stomodaeal and proctodaeal invaginations, a condition which could readily be interpreted as constituting a transitional stage between *Lepisma* and the higher pterygote insects. Recently Nusbaum and Fulinski (1906, 1909) have reinvestigated the origin of the mesenteron in two of the forms studied by Heymons, *Phyllodromia (Blatta)* and *Gryllotalpa,* and have obtained a different result, namely that the mesenteron is formed from the two ends and also from the median portion of the lower layer. Similar results were obtained by Hirschler (1906, 1909) in the Lepidoptera and Coleoptera. Heymons' conclusions and interpretations have also been contested by Escherisch (1900) in his paper on *Calliphora (Musca).* Carrière and Bürger's (1897) observations on the development of the mesenteron rudiments in the mason bee have already been mentioned (p. 71); Noack (1901) arrived at similar conclusions in the case of *Calliphora (Musca).* In contrast to the results of Heymons and all the other investigators of this subject Hammerschmidt (1910) finds that in the orthopteron *Dixippus* the mesenteron is formed exclusively from the median section of the lower layer, which in most insects produces the blood cells.

The conditions existing in the honey bee, as they have been described in this paper, obviously lend little support to the views of those who regard the mesenteron of insects as arising from the ectoderm of the stomodaeum and proctodaeum (III), since its rudiments are already formed long before the stomodaeal and proctodaeal invaginations appear; much less do they harmonize with the theory of the origin of the mesenteron from yolk cells (I). The relation of the mesenteron rudiments in the honey bee may be interpreted in either of two ways, and the one chosen will probably depend largely on the theoretical bias of the interpreter. First, the mesenteron rudiments may be referred to the mesoderm (II). Several facts can be cited in support of this view, for example: the continuity of the middle plate and the mesenteron rudiments during the earlier stages in their formation, and the fundamental similarity in the manner of formation of both the mesenteron rudiments and the mesoderm, all being formed by a contemporaneous inward migration of elements of the blastoderm. Second, the mesenteron rudiments may be considered, with Carrière (IV) as purely blastodermal in origin, since their manner of formation, although essentially similar to that of the mesoderm, differs much from it in detail, moreover the cells composing the mesenteron rudiments are from the first distinguishable from those of the mesoderm. A final decision between these two interpretations seems premature. The honey bee is a highly specialized member of a specialized order, and therefore an unsuitable form on which to base generalizations, since its development certainly presents many modifications of the type, moreover generalizations based on the study of one form are always unsafe.

It is sufficiently evident that in spite of the numerous papers dealing with the origin of the mesenteron in insect embryos, there is much need of further investigation, particularly of the more generalized types. Superficial study, however, would be worse than useless; the type of investigation demanded is the highest, requiring the delicate and precise methods of the cytologist, the best fixation and staining possible, a complete series of stages, a study of the origin of the rudiments cell by cell, and finally an eye single to the facts and regardless of preconceived theoretical considerations.

Before leaving the subject of the germ layers it will be necessary, in order to discuss the stages following, to describe briefly the structure of an embryo at the final stage of this period, Stage VII. This stage is illustrated by a series of transverse sections, represented by figures 29, 30, 31 and 32.

Figure 29 shows a section passing through the extreme anterior end of the embryo. The ectoderm here extends over about two-thirds of the circumference of the yolk and is much thickened in its lateral portions. Within the ectoderm is a crescentic mass of cells, the anterior mesenteron rudiment, which has grown both cephalad and laterad to form a cap-like mass covering the ventral side of the cephalic end of the yolk. In this and also in the next section the *amnion* (*Am*) is seen covering the exterior of the embryo as a thin sheet of flattened cells. This membrane will be discussed at length later.

Figure 30 also shows a section passing through the cephalic end of the embryo, intersecting it just caudad of the posterior limits of the anterior mesenteron rudiment. The opening in the ectoderm through which the anterior mesenteron rudiment reaches

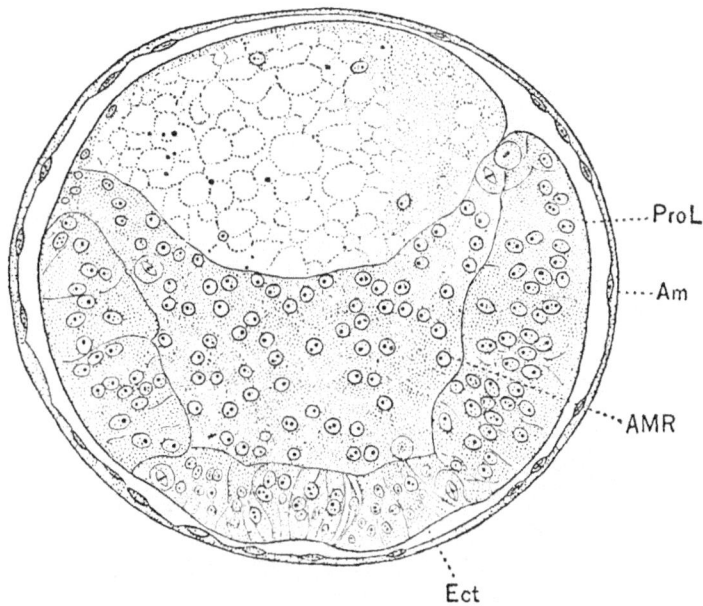

FIG. 29. Transverse section through the anterior end of an egg, Stage VII, passing in front of the rudiment of the stomodaeum, showing the procephalic lobes (*ProL*), anterior mesenteron rudiment (*AMR*), and amnion (*Am*), x 243.

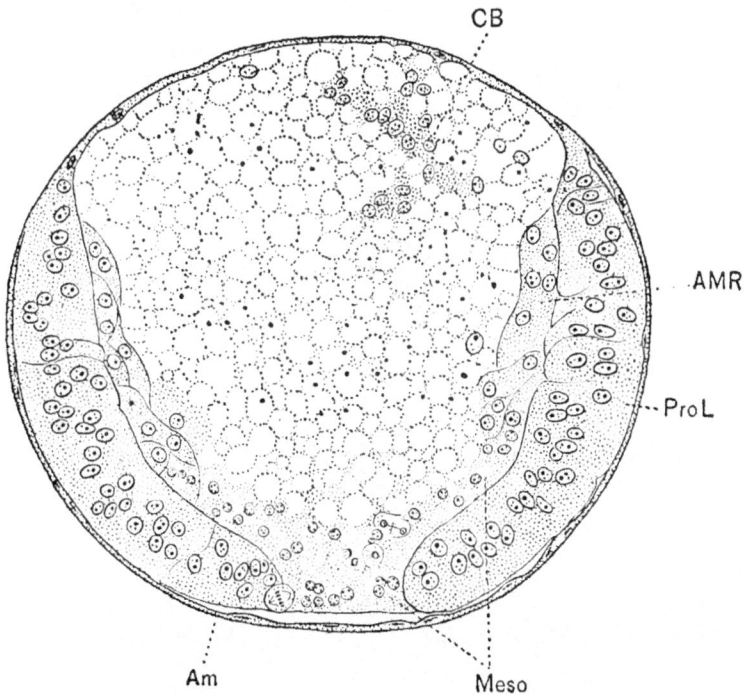

FIG. 30. Transverse section through the anterior end of an egg, Stage VII, passing just caudad of the rudiment of the stomodaeum, showing the procephalic lobes (*ProL*), anterior mesenteron rudiment (*AMR*), mesoderm (*Meso*) and amnion (*Am*), also the remains of the cephalo-dorsal body (*CB*), x 243.

the external surface is seen, but the cells filling it belong to the mesoderm (compare Fig. 26B), as the small size of their nuclei demonstrates.   The mesoderm (*Meso*) extends also a short distance laterad on both sides of the opening.   The anterior mesenteron rudiment (*AMR*) is represented in this section by a layer of cells two or three deep lying on each side against the inner surface of the lateral ectoderm, above the mesoderm.   The ectoderm has the same thickness and relative extent as in the preceding section.

Figure 31 represents a third section passing through the ceph-

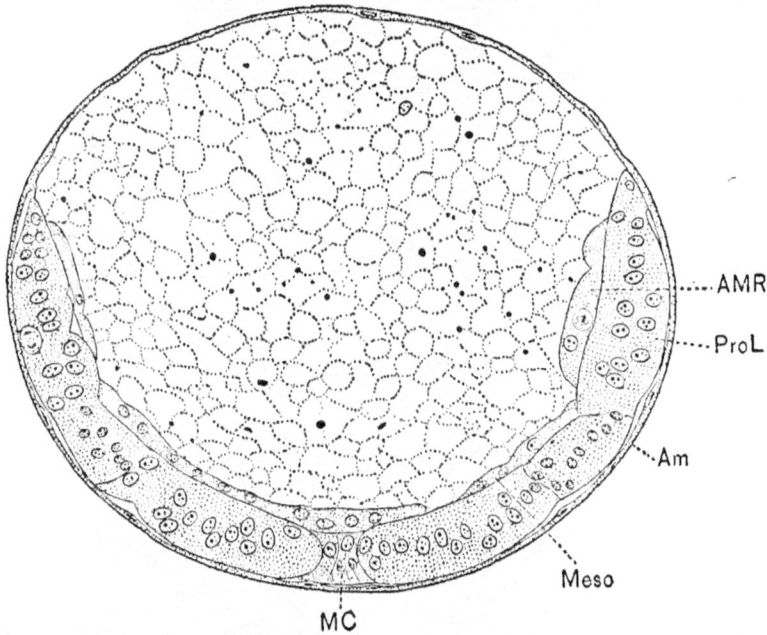

FIG. 31.   Transverse section through the anterior end of an egg.   Stage VII, intersecting the posterior margin of the procephalic lobes (*ProL*) The middle cord (*MC*) is shown, on each side of which is the neurogenous ectoderm from which the ventral cord is to be formed.   The mesoderm (*Meso*) the amnion (*Am*) and two or three cells of the anterior mesenteron rudiment (*AMR*) are also shown, x 243.

alic end of the embryo, a few sections caudad of the last.   The ectoderm here is seen to be diminishing slightly in both extent and thickness.   In the ventral mid-line is a strip of cells (*MC*) whose transverse section has somewhat the outline of an hour glass, and represents the point of juncture of the lateral folds. The significance and fate of this strip will be described in the section devoted to the nervous system.   Lining the ectoderm is a single layer of mesoderm, broken at two points.   High up, on the right hand side of the section, in place of mesoderm, two cells belonging to the anterior mesenteron rudiment are seen (*AMR*).

Figure 32 shows a section taken through the mid-region of the

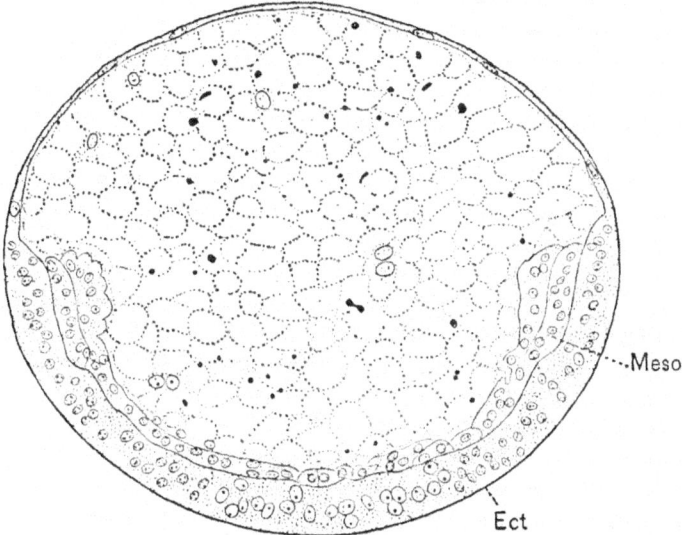

FIG. 32. Transverse section through the future thoracic region of an embryo, Stage VII-VIII, showing the form and relations of the ectoderm (*Ect*) and mesoderm (*Meso*), x 243.

trunk, and is representative of the condition existing throughout its extent, including the future thorax and abdomen. The ectoderm covers somewhat less than the ventral half of the egg, and is composed of a single layer of closely packed columnar cells. Within the ectoderm is the mesoderm (*Meso*), whose lateral extent is somewhat less than that of the ectoderm. The mesoderm, near the mid-line, is somewhat thinner than the ectoderm, but towards its lateral margins, which are rounded, its thickness approaches that of the ectoderm. It is clearly composed of two layers, continuous with one another at the lateral margins. Near the mid-line in its thinner portions the cells composing the two layers are flat and the line of separation between the layers is somewhat indistinct. Toward the lateral margins of the mesoderm, however, the cells composing the two layers become columnar, and the line of separation between the layers is sharp. This line of separation represents the cavity of the mesodermal somites.

A section through the caudal end of an embryo of Stage VII has already been represented in figure 28, and described in the part of this section relating to the posterior mesenteron rudiment.

VI

The Amnion and the Cephalo-dorsal Body

1. *The Amnion*

The cells destined to form the amnion, as described on p. 44, constitute the marginal portions of the ventral plate, and cover the dorso-lateral regions of the egg, bounding the median dorsal strip laterally (Figs. 19 and 24A, *Am*). At first the amnion-forming cells are not sharply demarcated from the remainder of the ventral plate, and seem to intergrade with the cells of the latter, but as the differentiation of the germ layers progresses the amnion cells also become differentiated, appearing more or less shortened or flattened as contrasted with the long prismatic form of the cells of the embryonic portion of the ventral plate. As soon as the amnion cells become distinguishable as such, it becomes evident that the amnion cells at the anterior end of the egg differ somewhat from those of the other regions. Over the entire cephalic end of the egg, the amnion cells are rounded in form, closely arranged and relatively numerous, in sections resembling a string of beads (Fig. 24A and 26A, *Am*). The characteristic transparency of the inner ends of the amnion cells, noted by Petrunkewitsch (1903) is especially evident. Caudad of the cephalic region the amnion cells become relatively fewer and their form also becomes more flattened (Figs. 20A and B and 21, *Am*). These differences are, however, merely temporary and disappear as the amnion increases in extent. Prior to Stage V the amnion therefore consists of two longitudinal bands of epithelial cells, separated by the median dorsal strip and joining the ventral plate laterad. These amniotic bands are widest at their anterior ends, which cover the cephalic end of the egg, and their component cells are here more numerous and less flattened than elsewhere, as just mentioned. At Stage V the two bands begin gradually to widen, the cells of their inner margins creeping up over the dorsal strip, which becomes submerged in the yolk (Figs. 20A and B, 21, 24B, 34B and C, *Am*). Subsequently then is a fusion of the amniotic bands along the dorsal mid-line of the egg, beginning first at the cephalic end, and occurring somewhat later over the remaining extent of the egg, being completed at Stage VI.

While the amnion is thus covering the dorsal side of the yolk it commences also to cover the ventral side. This process can best be observed in fresh material, in which the outlines of the amnion are beautifully clear. In fixed material, whether examined in alcohol or stained and cleared, the outlines of the amnion are frequently invisible, except in actual or optical sections. At Stage V or a little earlier, on the ventral side of the egg, the amnion together with the anterior end of the germ band separate from the yolk (Fig. 26A). This is not primarily due to the depression of the yolk caused by the development of the anterior mesenteron rudiment as the figure might suggest, since the separation is already evident when this rudiment is in its earliest stages; it is probably first brought about by an increase in the superficial extent of the amnion in this region, as indicated by its convex or arched form. Soon after, the amnion separates from the yolk over the entire cephalic pole of the egg, rising up in the shape of a hemispherical cap (Fig. V). Next—at Stage VI—it severs its connection with the germ band around the anterior end of the latter and slides over it in the form of a hood or cowl, thus forming the *cephalic fold* (Fig. 32, *1am*, and 26B *Am*). This separation of the amnion from the germ band progresses caudad along its lateral margins, accompanied by the caudad extension of the cephalic fold over the ventral face of the germ band in such a way that the free edge of the cephalic fold forms a semicircular curve with its concave side directed caudad (Fig. 32, A-C, *1am*). When the cephalic fold has covered about one-half of the ventral face of the embryo, a second amniotic fold, the *caudal fold* (Fig. 32 B, *2am*), appears at the extreme caudal end of the germ band. This fold is formed like the head fold, by separation of the amnion from the caudal end of the germ band, and first appears as a crescentic membrane. Since the caudal end of the germ band is now curved completely around the caudal pole of the egg, so that the former lies on the dorsal side of the egg, the caudal fold also lies on the dorsal side of the egg with its concave edge directed toward the caudal pole. The caudal fold increases in extent in the same manner as the cephalic fold, progressing slowly toward the caudal pole of the egg, the lines of rupture of the caudal fold and the germ band extending to meet those of the cephalic fold (Fig. 32C). The two folds thus approach one another, the cephalic fold moving at a much more rapid rate than its counterpart; soon the two meet and fuse near the caudal pole of the egg (Fig. 32C and D). This occurs slightly prior to Stage VIII. While the amnion is thus covering the ventral side of the egg, and consequent to its severance from the edges of the germ band, it also separates from the yolk on the dorsal side of the egg. This separation is directly connected with the separation of the amnion from the edges of the germ band, since it must at the same time also separate from the yolk at this point. The separation thus initiated is continued dorsad, the amnion being, so to speak, peeled off from the yolk, which is thus left bare except for the thin protoplasmic pellicle surrounding it (cf. Figs. 29 and 30). When the formation of the amnion is completed, or shortly afterwards, the amnion therefore forms a complete envelope surrounding the embryo, and free from the embryo at the two ends of the egg, but closely applied to the embryo elsewhere. Its outline is similar to that of the chorion, except that it is shorter, leaving a considerable space vacant between amnion and chorion at the ends of the egg. Since the space surrounding the egg, between the latter and the chorion, is filled with a watery fluid, it follows that the space between the embryo and amnion is also filled by this same fluid.

The cephalic fold of the amnion, as described above, is at first composed of cells which are rounded in form (Figs. 24B, 26A). As this fold progresses over the surface of the germ band its cells become gradually thinner and flatter (Figs. 26B, 29, 30 and 31, *Am*). At its completion, at Stage VIII, its average thickness is scarcely greater than that of the chorion. At the ends of the egg the amnion is somewhat thicker than elsewhere, and in these regions the nuclei are oval in outline, forming lenticular swellings (Fig. 29). Elsewhere, over the body of the embryo, the amnion is scarcely thicker than the chorion, its nuclei being flattened to such an extent that in sections they frequently appear as short dark lines (Fig. 30). Comparing the extreme tenuity of the amnion, when completed, and the superficial extent of each of its component cells as compared with the thickness and slight superficial extent of the amnion cells, at earlier stages, particularly those of the head fold, it is apparent that the development of the amnion is due principally, if not exclusively, to a mere extension or spreading out of the original cells present at least as early as Stage IV. This was essentially the view taken by both Bütschli (1870) and Grassi (1884). Neither of these investigators saw any division of the amnion cells, nor has the writer observed them.

Bütschli (1870), Kowalevski (1871) and Grassi (1884) have described and figured the formation of the amnion in the honey bee. Bütschli's account, based exclusively on observations of fresh eggs, is full and substantially correct, recording, among other details, the covering of the dorsal surface of the egg by the amnion cells. Kowalevski's account is less extended than that of Bütschli, and while correct as regards the topographical relations of amnion and germ band, erroneously describes the amnion as originally composed of two layers, which subsequently fuse to form one. This error was possibly due to Kowalevski's contemporaneous studies on *Hydrophilus,* in which, as is well known, there are two embryonic membranes. Grassi's account, although based in part on actual sections, adds but little to that of Bütschli. Grassi incorrectly describes the meeting of the cephalic and caudal folds as taking place on the ventral surface of the egg midway of its length, whereas it normally takes place at or near the caudal pole of the egg.

Before entering upon a comparison of the embryonic envelope (amnion) of the bee with the embryonic envelopes of the other pterygote insects, it will not be out of place to recall the manner in which these are formed. Briefly stated, it consists essentially in the elevation, around the embryonic rudiment, of a fold of the extra-embryonic blastoderm, which then extends over the embryo from all sides, its edges finally meeting and fusing (see Korschelt and Heider, Fig. 133). Contemporaneous with this fusion is the separation of the two layers composing the fold, so that

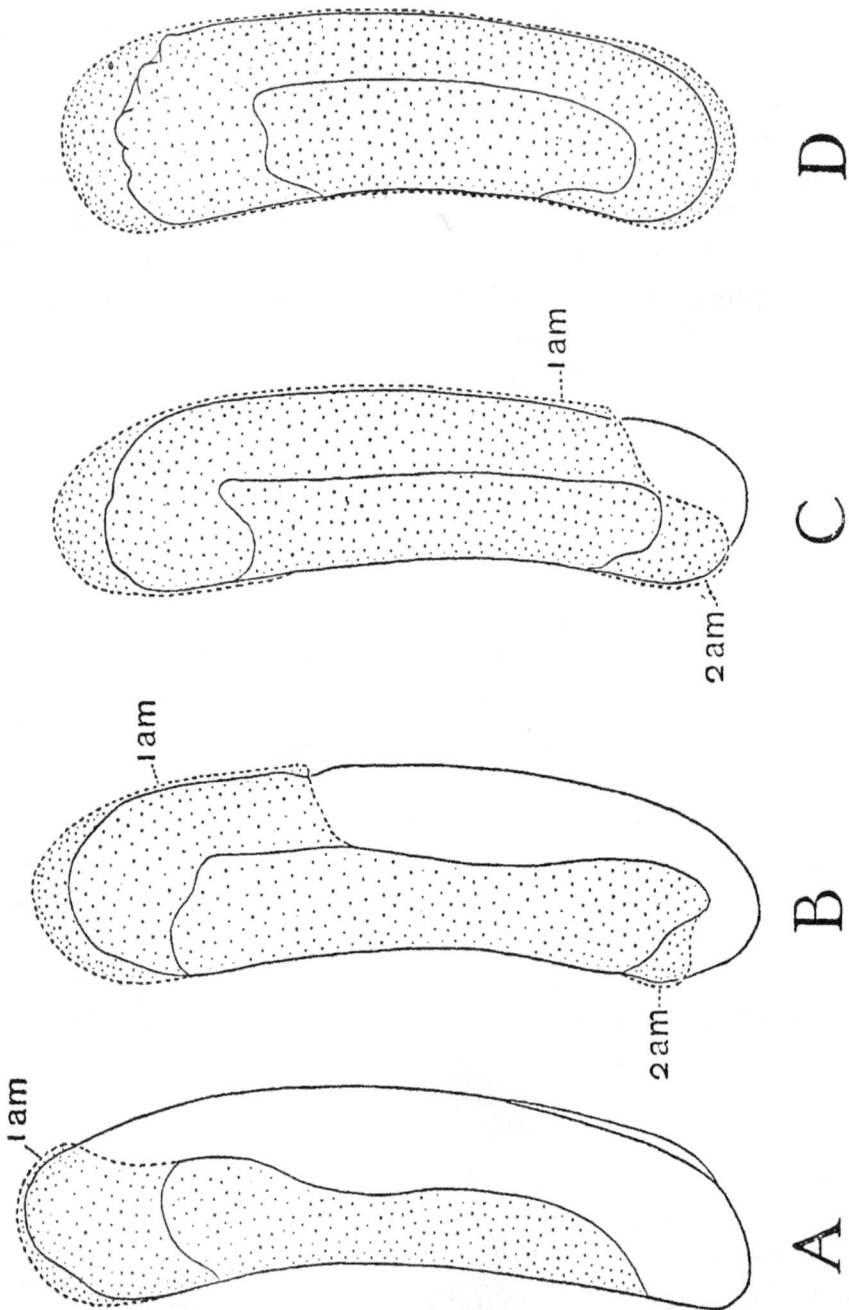

FIG. 33. Side views of four embryos, drawn in diagrammatic form, showing four stages in the development of the amnion. Outlines drawn with a camera lucida from living eggs. The amnion (*1Am, 2Am*) is stippled, with dotted outline, except when it adheres to the embryo or yolk. In B, C and D the amnion is represented for the sake of clearness, as standing off from the embryo further than is actually the case. A corresponds to Stage VI, B to Stage VII, C to Stage VII-VIII, D to Stage VIII-IX, x 41.

the embryo thus becomes covered by two separate layers, one enclosing the entire egg, the *serosa,* and one covering only the embryo, the *amnion.* The amniotic fold, the common rudiment of the amnion and serosa, does not usually develop simultaneously around the entire margin of the embryo, but appears commonly first at its anterior and posterior ends, thus forming a cephalic and a caudal fold. These conditions may be considered typical of the majority of pterygote insects. There are, however, some exceptions of which the honey bee is an example. Here, as is already apparent, but one envelope is present, which, since it covers the entire egg inclusive of the germ band, is therefore comparable to the serosa rather than to the amnion. This is the opinion held by Heider (1891). The term amnion nevertheless has been applied by both Bütschli and Grassi to the single embryonic envelope of the honey bee, and it therefore seemed inadvisable to change it.

In addition to the honey bee, embryonic envelopes have been observed in the following representatives of the non-parasitic Hymenoptera: *Formica, Myrmica* (Ganin 1869), *Polistes, Formica* (Graber 1888), *Hylotoma* (Graber 1890), *Chalicodoma. Polistes* (Carrière 1890, and Carrière and Bürger 1897), and *Camponotus* (Tanquary 1913). Graber states that two embryonic membrances were found in all the three forms studied by him. On the contrary Carrière and Bürger (1897) explicitly state that in one of the forms studied by Graber, *Polistes,* only one embryonic membrane is present. This is also the case in *Chalicodoma.* In the ants but one membrane is found, according to both Ganin (1869) and Tanquary (1913).

In some of the parasitic Hymenoptera, for example *Biorhiza, Rhodites* (Weismann 1882) *Platygaster* (Kulagin 1897), an embryonic membrane is formed which is also single, but the formation of this—at least in the case of *Platygaster,*—is so peculiar and so different from that found in other insects that its homology with the embryonic membrane of non-parasitic insects is perhaps open to question.

Concerning the fate of the amnion in the honey bee the only data given by previous observers are those of Bütschli (1870) who says (pp. 533-534): "This envelope persists during the entire development of the embryo, and like the egg envelopes is finally torn by the active movement of the young larva." The writer has not observed this process in the living larva, but a study of sections confirms Bütschli's statements. Up to Stage XIV the amnion is intact, but at Stage XV, when the young larva has become flexed ventrad, and has ruptured the chorion, the amniotic membrane is evident only in fragments, usually clinging to the chorion. These fragments or shreds are very much thicker than was the amnion before its rupture, and always contain a number of ovoid nuclei, closely grouped together, indicating that the fragments had contracted and suggesting that the amnion had been under tension previous to its rupture. In *Chalicodoma*, according to Carrière (1890, 1897) some time before the hatching of the larva the embryonic envelope becomes torn into fragments, which become thickened to form narrow twisted bands of polygonal cells. These disappear before hatching, being apparently absorbed. In the pterygote insects of other orders,—Orthoptera, Coleoptera—the later history of the embryonic envelopes is very varied, and usually associated with a shifting in the position of the embryo relative to the yolk (*blastokinesis*, Wheeler 1893). The fate of these envelopes, briefly stated, is as follows: The serosa is either cast off or absorbed by the egg; the amnion is usually also absorbed, but in some cases has been described as contributing to the formation of the dorsal body wall, as in *Meloë* (Nusbaum 1890).

2. *The Cephalo-dorsal Body*

Associated with the formation of the amnion and contemporaneous with that of the germ layers, is a peculiar structure to which the writer (1912) gave the name "cephalo-dorsal disk." This structure (Fig. 34 A-C), which attains its maximum size about Stage V, is extremely variable in location, form and size. On account of its variability in size it would be better to term this structure the "*cephalo-dorsal body.*" It is situated on the dorsal side of the egg usually near the cephalic pole and opposite to the anterior mesenteron rudiment (Fig. 34A), but in one egg, which appeared to be normal in every other respect, the cephalo-dorsal body was found situated almost precisely at the cephalic pole itself. The form of this body, while very variable, usually ap-

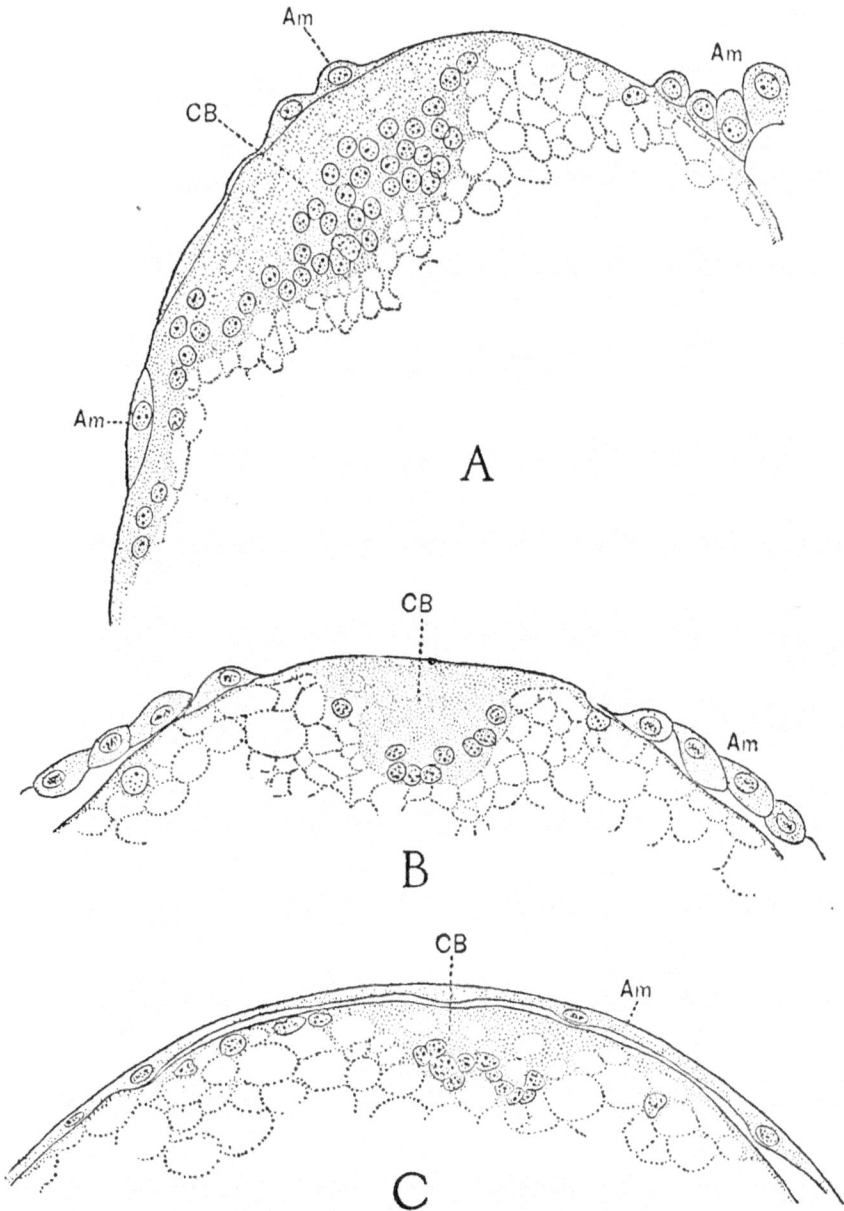

FIG. 34. Parts of sections showing the cephalo-dorsal body. A, cephalic part of a median sagittal section of an egg, Stage V, showing the cephalo-dorsal body at the time when it is best developed and most conspicuous. B, dorsal part of a transverse section of an egg, Stage V, showing the amnion (*Am*) advancing over the cephalo-dorsal body toward the dorsal mid-line. C, the cephalo-dorsal body covered by the amnion, x 387.

proaches that of an ellipsoid, flattened dorso-ventrally, with its long axis lying in the sagittal plane. In many cases in its form and longitudinal extent it approximates the one illustrated in figure 34A, in some it is smaller, while in two cases the cephalo-dorsal body was tongue or club-shaped, its smaller end extending back in the dorsal mid-line for approximately one-third of the entire length of the egg. Its external surface, in the earlier stages forms a part of the external surface of the egg, in later stages it is partly or entirely covered by amnion cells. Its inner surface, usually strongly convex, is indented by the alveolar spaces representing the yolk spheres, while extending out between the indentations are delicate protoplasmic processes continuous with the interstitial protoplasmic meshwork (Fig. 34A and C). The cephalo-dorsal body is therefore, like the dorsal strip, closely associated with the yolk. In structure the cephalo-dorsal body appears to be a syncitium, composed of rather clear and somewhat vacuolated cytoplasm, within the inner half of which numerous nuclei are embedded. At its posterior edge it becomes continuous with the dorsal strip, as shown in the figure (Fig. 34A).

In origin the cephalo-dorsal body appears to be little more than a localised swelling of the dorsal strip. At Stage IV and the stages intervening between Stages IV and V the cephalo-dorsal body, as compared with its condition at Stage V, is located somewhat closer to the cephalic pole, is flatter and contains fewer nuclei, being plainly nothing more than the slightly thickened anterior end of the dorsal strip. At about Stage V it increases rapidly in size and in the number of nuclei contained within it, until it attains its maximum size, as shown in the figure, but its connection with the dorsal strip is never lost and is always readily apparent. Moreover, in many cases smaller but similar swellings of the dorsal strip exist at various points on the dorsal mid-line in the anterior half of the egg. One of these is illustrated in figure 20A.

The rapid increase of the cephalo-dorsal body in size and more particularly in the number of nuclei contained within it, is difficult to explain. While it is not impossible that it owes its origin in part to cells derived from the yolk, there is no satisfactory evidence that this is the case. The nuclei seen in the yolk during the formation of this body are usually larger and rounder than those contained within the latter, moreover they are not present in sufficient numbers in the immediate neighborhood of the cephalo-dorsal body to make it seem probable that they contribute largely to its formation or growth. It is also possible that the number of nuclei may be increased by either direct division or fragmentation of the nuclei originally present, but this could not be demonstrated to be the case, although in many instances this was suggested by the appearance of the nuclei. The latter are so small and so frequently crowded closely together that it was found impossible satisfactorily to decide this question. It is, however, certain that the nuclei differ among themselves in size, and that they are also very frequently lobed or constricted.

Soon after reaching its maximum size the cephalo-dorsal body during Stage VI breaks up into amoeboid cells or small syncitia containing one to several nuclei. Some of these wander ventrad towards the center of the egg (Fig. 30), but many remain near the dorsal surface of the yolk. At Stage VII the remnants of the cephalo-dorsal body are still recognizable at the cephalic end of the egg as irregular branching islands of rather pale granular cytoplasm enclosing one or more small nuclei. The outlines of the nuclei are always irregular and often faint, and when grouped closely together as is frequently the case, many of the nuclei appear to be in process of fusing with one another (Fig. 34C). At Stage VIII the remains of the cephalo-dorsal body are evident, lying close beneath the anterior mesenteron rudiment, near its posterior end (Fig. 87A,CB). From this point they travel toward the caudal pole of the egg as if drawn by the rudiment until they reach the point of junction of the anterior and posterior mesenteron rudiments. During this period the vestiges of the cephalo-dorsal body usually present the appearance of a thin layer of slightly yellowish protoplasm, containing a few nuclei which are situated near its inner surface, as shown in figure 87A, CB. In some preparations, however, the cephalo-dorsal body seems to have remained undiminished in size up to Stage VIII or IX. It is then quite conspicuous and plainly evident even at a low magnification. In these cases the appearance is almost precisely like that figured by Petrunkewitsch (1902, Fig. 8) for the drone egg,

the cephalo-dorsal body being here represented by the "*Rz*" cells.[15]
At Stage X the vestiges of the cephalo-dorsal body may still be
seen lying beneath the dorsal wall of the mesenteron, but they
soon become indistinguishable from the elements of the yolk, and
unquestionably like them suffer ultimate absorption. The cephalo-
dorsal body was first seen by Grassi (1884), who has represented
it in figure 1 of Plate X, but did not mention it in the text. Dickel
(1903) was the first to describe this body, which he named the
"yolk plug," since he regarded it as derived from yolk cells which
migrated to its point of origin, this point being designated as the
blastopore. The "yolk plug" was regarded by Dickel as the anterior mesenteron rudiment. This assumption was founded on the
circumstance that the "yolk plug" had disappeared at the time
when the anterior mesenteron rudiment came into prominence.
It will hardly be necessary to say that this assumption is wholly
wrong, and must have been due to defective observation, since
any series of sections showing the "yolk plug" should also show
the true anterior mesenteron rudiment, since in fact the "yolk
plug" becomes evident only after the anterior mesenteron rudiment has begun to be formed. Moreover, the remnants of the
cephalo-dorsal body ("yolk plug") are plainly evident long after
the anterior mesenteron rudiment is fully formed.

Petrunkewitsch (1902), in a paper on the fate of the polar
bodies in the drone egg, describes and figures a body formed in
the cephalo-dorsal region of the egg by cells—termed for brevity
"*Rz*-cells"—which owe their origin to a single cell formed by the
union of the second polar body with the inner half of the first.
This structure formed by the "*Rz*-cells" apparently agrees closely
in form, structure and position with the cephalo-dorsal body of the
worker egg. Petrunkewitsch, however, states that he did not
find this body in worker eggs, and gives a figure to illustrate this
statement, but this figure (4) is obviously of an earlier stage than
those in which is found either the cephalo-dorsal body or that
formed by the "*Rz*-cells." The fate of the latter according to
Petrunkewitsch's account, is as follows: It first sinks below the

[15] Petrunkewitsch (*loc. cit.*) states that mitotic figures were found in
this cell mass. The writer has never seen them in the cephalo-dorsal body
at any stage.

surface into the yolk and becomes covered by blastoderm (amnion). So far its history corresponds to that of the cephalo-dorsal body. Next the "Rz-cells" glide caudad along the dorsal side of the yolk until they reach the middle of the abdomen, increasing in number meanwhile by mitotic division. Here they are overtaken by the advancing ectoderm and the epithelium of the mesenteron and come to lie between these two layers. Finally the "Rz-cells" divide into two groups, one on each side of the embryo. These then migrate ventrad, enter the coelomic cavities on each side of the third, fourth and fifth abdominal segments and constitute the male sex cells.

Nachtsheim has already (1913) explicitly affirmed the identity of the "Rz-cells" of Petrunkewitsch and the "yolk plug," of Dickel solely on the evidence afforded by the two papers.

The writer has not so far studied the "Rz" cells in the drone egg, so that a final conclusion regarding the identity of the cephalo-dorsal body and the "Rz" cells may seem premature. Nevertheless, a comparison of Petrunkewitsch's figure with the corresponding stages in the worker egg, as well as with the figures given by Dickel (1904) leads one almost inevitably to the conclusion that the "Rz" cells, the "yolk plug," and the "cephalo-dorsal" body are identical.

The significance and possible homologies of the cephalo-dorsal body are extremely uncertain. Its rapid disappearance and great variability suggest that it is a vestigial structure. Hirschler (1909) has described, in the chrysomelid beetle *Donacia*, a structure termed by him "the primary dorsal organ," which in several respects resembles the cephalo-dorsal body of the honey bee. This "primary dorsal organ" consists of an oval area of the blastoderm lying in the dorsal mid-line, which sinks down into the yolk and becomes covered over by blastoderm, its component cells meanwhile sending out processes into the yolk and also showing evidences of degeneration. Finally "the primary dorsal organ" degenerates completely and is absorbed by the yolk. In position, overgrowth by blastoderm and final absorption by the yolk this structure corresponds quite closely to the cephalo-dorsal body of the honey bee, but in one respect it differs: in the bee the cephalo-dorsal body is formed during the formation of the germ layers, while in *Donacia* the "primary dorsal organ" is formed prior to the formation of the germ layers. In respect to other insects little can be said that is not pure conjecture, since the data afforded for comparison are too scant. To those familiar with insect embryology comparisons with "the micropylar organ" (Wheeler 1893) or "procephalic organ" (Claypole 1898) of *Anurida* and the "indusium" of *Xiphidium* (Wheeler 1893) will suggest themselves, but only further investigations on the embryology of insects with special regard to the question as to whether a vestigial "dorsal organ" is of frequent occurrence in insect embryos can be of real value in determining the homology of these various anomalous structures.

VII

GENERAL ACCOUNT OF THE DEVELOPMENT OF THE EMBRYO WITH ESPECIAL REFERENCE TO THE EXTERNAL FORM

1. *Changes in the Form of the Egg up to Stage VIII*

The developmental changes undergone by the egg, from the beginning of the cleavage to the close of the formation of the germ layers, have already been described in detail, nevertheless there is one phase of the development which so far has been left out of account, that is, the changes in the external form of the egg. As far as Stage VIII the original outlines of the egg are in a large measure preserved; these have been broken only by the amniotic folds, and in a slighter degree, by the lateral folds. On the other hand the proportions of the egg have been somewhat altered; during the stages described, it has appreciably shortened, as may be seen by comparing figures I and VII. This shortening is very evident in the actual specimens, in which the chorion is intact, since at Stages VII-VIII there is a space of considerable size left vacant between the embryo and the chorion at the two ends of the egg (Fig. 35D). Examination of early stages (I) shows that at the beginning of cleavage the egg contracts slightly in the long axis, as shown by the appearance of a narrow space between the egg and the chorion at the anterior end and frequently at both ends. The amount of contraction does not usually exceed about 5 per cent of the original length of the egg, as estimated by the length of the chorion (Fig. 35A). This shortening of the egg in its long axis was seen in the living egg by both Bütschli (1870) and Kowalevski (1871) and mentioned by them as being one of the earliest phenomena of development. No further shortening of the egg takes place until the time of the formation of the germ layers, but during this latter period (Stages IV-VII) the egg again shortens rapidly until at Stage VIII the egg (embryo) is now from 20 to 25 per cent shorter than the chorion, relatively large spaces, filled with fluid, being left vacant at both ends of the egg, between the embryo and the chorion (Fig. 35B

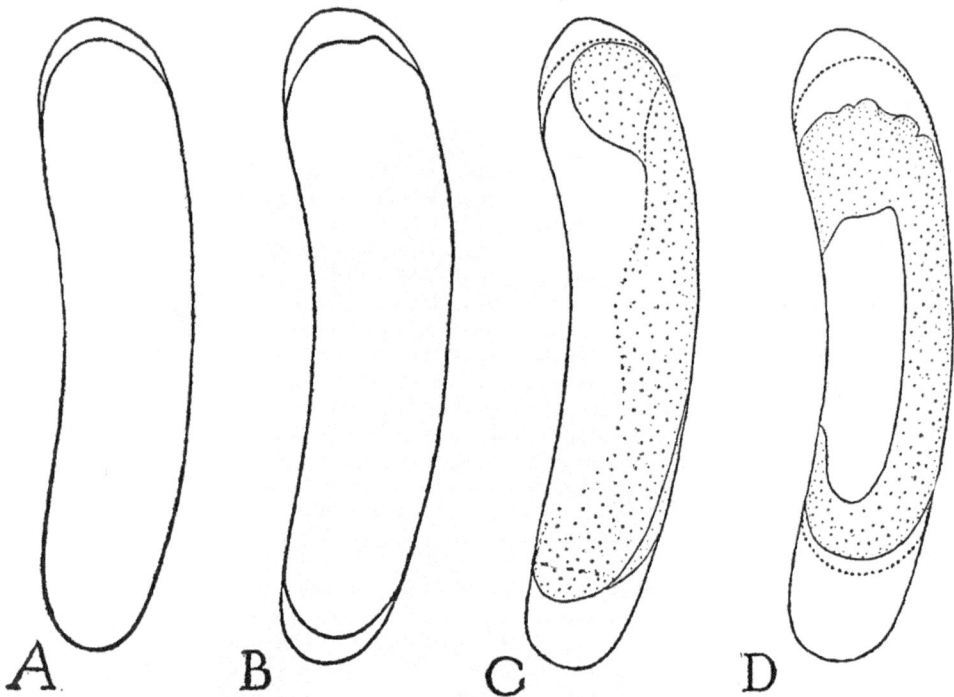

Fɪɢ. 35. Profile drawings of eggs showing four stages in development, to illustrate the relative dimensions of the egg and chorion. Outlines drawn with camera lucida from fixed material. Amnion represented by dotted line. A, cleavage and blastoderm stages; B, Stage IV-V; C, Stage VI; D, Stage VIII, x 41.

and C). These spaces are invaded by the amnion. A comparison of figures 35B and D shows that this second shortening of the egg is not only contemporaneous with the formation of the germ layers, but that it is also accompanied by the curving of the germ band around the ends of the egg. It is therefore evident that the shortening which takes place has principally affected the yolk, and furthermore that the germ band itself has actually lengthened but little, if at all, between Stages IV and VIII. The embryo does not subsequently increase materially in length until immediately before hatching, since eggs 70-73 hours old still show spaces of considerable size at their ends, between the embryo and the chorion. At the time of hatching, however, the young larva, after rupture of the chorion, increases greatly in length, as shown in figure XV.

These changes in the length of the egg, with the exception of the initial shortening, are not explicitly mentioned by any of the previous investigators; nevertheless, the figures of both Bütschli

(1870) and Kowalevski (1871), which represent the embryos of different stages surrounded by the chorion, show that they observed these changes precisely as they are described above.

2. *The Development of the Embryo*

In the study of any developmental processes a knowledge of the end stage is necessary to an intelligent comprehension of the earlier stages, since these are only isolated examples chosen from what is in reality a continuous process leading to the end stage. This is, in a sense, their goal, but in a larger sense is itself only a more or less arbitrarily chosen stage in the continuous process of the life cycle. The end stage of the development in the egg is of course attained when the embryo breaks the egg envelope or chorion and becomes a larva. This stage, Stage XV, reached in from 74 to 76 hours after the egg is laid, is illustrated by figure XV, representing a recently hatched larva. This larva is drawn not from a living specimen, but (like the others of this series) from one which has been stained, cleared and mounted. The body wall is represented as much more transparent than it is in reality, in order to show more clearly the organs contained within, while the sharpness of outline of these is exaggerated.

The larva as a whole is long cylindrical in form, slightly bent toward the ventral side, like a bow, and tapers gently toward the caudal extremity, which is obliquely truncate. The larva is divided by constrictions into a short and rounded head and thirteen segments, the latter being approximately equal in length and collectively constituting the trunk. On the dorsal side the constrictions are narrower and deeper than on the ventral side. The head is of the same diameter as the anterior part of the trunk, and the constriction separating the two is only a trifle deeper and broader than those separating the trunk segments. On the ventral side of the head is the mouth (*Mth*). In front of the latter and directed downwards is a blunt rounded lobe, the *labrum* (*Lm*), while bounding the mouth laterally are two pairs of rather large papilliform processes, the *mandibles* (*Md*) and the first maxillae (*1.Mx*). At the rear the mouth is bounded by a single flat lobe, the *labium* (*Lb*).

The alimentary canal is simple, consisting of a tubular oesophagus (*Oe*) which extends scarcely beyond the neck and opens into the *mid-intestine* (*MInt*). This is a capacious cylindrical sack which occupies the greater part of the space contained within the trunk, extending caudad to the twelfth segment, where it terminates blindly in a rounded end. It is succeeded by the short *hind-intestine* (*HInt*) whose inner (cephalic) end is also blind, but whose caudal end opens to the exterior by an opening on the obliquely truncated caudal face of the terminal or thirteenth segment. Opening into the cephalic end of the hind-intestine are two pairs of slender tubes, the *Malpighian tubules* (*Mal*), which lie against the lateral faces of the mid-intestine and pursue a sinuous course cephalad to the fifth trunk segment, where they end blindly. The function of the Malpighian tubules is supposed to be excretory.

Traversing almost the entire length of the trunk and lying near the ventral body wall are a pair of straight tubes, less slender and with somewhat thicker walls than the Malpighian tubules, the *silk glands* (*SlkGl*). Their caudal ends are blind, at their cephalic ends they contract somewhat, their walls at the same time becoming thinner, they then unite in the neck region to form a single median thick walled duct which opens at the end of the labium.

The nervous system comprises a *brain* (*Br*) and a *ventral nerve cord* (*VNC*). The brain is a large mass situated in the dorsal half of the head, set astride, as it were, of the oesophagus. It consists of two lateral lobes united together in the mid-line by a slender commissure. Each lobe contracts ventrad to a slender stalk which joins with the anterior end of the ventral cord below the oesophagus. The ventral cord consists of 12 successive swellings or *ganglia*. These, when looked at from above or below are seen to be double, and in fact each represents a pair of ganglia fused in the mid-line (Figs. XIII and XIV). Each of the pair is connected with the ganglia of the same side in the immediately adjacent segments by a *connective*, thus the entire cord appears as though pierced by a series of intersegmental round openings, and therefore somewhat resembles a ladder in form (Figs. XIII and XIV). The first ganglion is much thicker and longer than any of the others, this is the *suboesophageal ganglion* (*SoeGng*) and is a compound ganglion composed of the three pairs of ganglia belonging to the three segments represented by the mandibles, and the first and second maxillae, respectively. The last or twelfth ganglion is also compound, probably also representing the fused ganglia of three segments.

The respiratory system consists essentially of ten pairs of *spiracles* (*Sp*) and two *longitudinal tracheal trunks* (*LTraT*), one on each side. The spiracles are arranged in two longitudinal rows, one on each side of the larva, each trunk segment, from the second to the eleventh inclusive, bearing a single pair. The spiracles consist of plate-like thickenings of chitin. In the center of each of these is an opening leading into a narrow and short tubular branch which in turn connects with the longitudinal tracheal trunk of each side. These traverse the entire length of the body, lying between the mid-intestine and the body wall. At their anterior ends each tracheal trunk bends around the anterior end of the mid-intestine above the oesophagus to join its counterpart of the opposite side, thus forming a closed loop. At the posterior end of the mid-intestine the posterior ends of the longitudinal trunks form a similar loop, which is situated below the hind-intestine. The longitudinal tracheal trunks give off both dorsal and ventral branches which arise near those leading to the spiracles. The larger ventral branches join with the corresponding branches of the opposite side to form a series of tracheal loops or commissures, one of which is found in each trunk segment, from the first to the eleventh inclusive. The dorsal branches split up into finer branches which are distributed to the dorsal half of each segment.

The *heart* (*Ht*) is a thin walled tube, slightly constricted at the boundaries between the segments, and extending the length of the trunk in the dorsal mid-line, above the mid-intestine.

The rudiments of the *ovaries* (*Ov*) are seen as somewhat flattened elongated bodies lying above the mid-intestine in the dorso-lateral region of the seventh and eighth trunk segments and extending half way into the ninth segment.

After having briefly outlined the structure of the newly hatched larva, the account of the development will be resumed with Stage VIII, which follows immediately after the completion of the germ layers (Stage VII).

Stage VIII, 44-46 hours (Figs. VIIIa and b). In comparison with the previous stage the germ band has increased slightly in breadth and also in length. At Stage VII its anterior end barely

reached the cephalic pole, and was expanded laterally to form
a pair of flat lobes, the *procephalic lobes,* rounded on their anterior
and lateral margins. At Stage VIII the germ band has lengthened
and its anterior end has now moved around the anterior pole of
the egg so that the procephalic lobes come to lie on its dorsal side
(Fig. VIIIb). They are now no longer simple flat expansions of
the germ band, but are subdivided into two successive pairs of
lobes, an anterior and a posterior pair. The anterior pair, the
protocerebral lobes (*1Br*), are the larger, and in form and position may be compared to a saddle, placed on the dorsal side of the
anterior end of the egg. Their lateral ends are rounded and
extend ventrad half way down the lateral surfaces of the egg.
The second pair, the *deutocerebral lobes* (*2Br*), which with reference to the embryo, lie caudad of the first, are situated just a
trifle dorsad of the cephalic pole of the egg, and are also saddle
shaped but are somewhat smaller, and their lateral margins,
instead of being broadly rounded converge on each side
toward small rounded elevations; these elevations are the rudiments of the *antennae* (*Ant*). In the mid-line of the embryo,
about half way between the protocerebral and deutocerebral lobes,
just dorsal to the cephalic pole of the egg, is a low elevation,
broad at its anterior (dorsal) end, narrow and cristate at its
posterior (ventral) end. This is the rudiment of the *labrum*
(*Lm*). Just below (caudad of) the labrum, at the cephalic pole
of the egg, is a shallow circular depression, the *stomodaeum*, the
common rudiment of the mouth and oesophagus. Immediately
behind the mouth and near the mid-line are a pair of low rounded
swellings, the *tritocerebral lobes* (*3Br*). Following these, but
nearer to the lateral margins of the germ band and in line with
the antennal rudiments are three successive pairs of low rounded
swellings. The first and largest of these are the rudiments of the
mandibles (*Md*), the next those of the *first maxillae* (*1Mx*), and
the last those of the *second maxillae* (*2Mx*). These appendages
with the segments to which they belong, complete the head of the
future larva and imago; the remainder of the germ band corresponds to the trunk of the larva, and the thorax and abdomen of
the imago. On this portion of the germ band, destined to form
the trunk, eleven segments are plainly indicated in side view (Fig.
VIIIb) by narrow intersegmental light zones alternating with

darker ones; the last two segments, the twelfth and thirteenth, are however not yet clearly separated. On the ventral side the intersegmental light zones broaden out and unite along the mid-line, so that here the segments are represented by dark processes stretching out from the sides toward the mid-line. These darker areas represent the mesodermal somites.

The germ band is of almost uniform width from the mandibular segment to the twelfth trunk segment, widening only a trifle at the fourth and fifth trunk segments. On the three segments next following the head, on the ventral side are seen the faint outlines of three pairs of larger appendages, which are the rudiments of the *thoracic legs* (*1L, 2L, 3L*).

Contemporaneous with the appearance of the rudiments of the mouth parts and legs is the appearance of the invaginations which form the silk glands and tracheae. The rudiments of the silk glands (*SlkGl*) are plainly seen as a pair of small pale apertures transversely elongated, situated just caudad of the bases of the rudiments of the second maxillae. The common rudiments of the spiracles and tracheal system appear as a linear series of irregular apertures on each side of the germ band, a pair being evident on the second, third, fourth, fifth and sixth trunk segments, the first pair being the largest and most distinct, and the last the smallest and faintest. Since the definitive number, ten pairs, is found in the stage next following, it is apparent that the tracheal rudiments at Stage VIII are just making their appearance and also that they develop successively caudad.

Stage IX, 52-54 hours (Fig. IX). At this stage the germ band has not altered in its general appearance or proportions but all of the rudiments present in the preceding stage are more sharply defined. Those of the mandibles and maxillae are now well defined ovoid swellings, distinctly marked off from one another and plainly raised above the general level of the germ band. The premandibular appendages or second antennae are at this stage best developed, being well defined rounded eminences situated just in front of the mandibles and a trifle nearer the mid-line. A line connecting them would pass just below the stomodaeal invagination. The rudiments of the first pair of thoracic legs are more clearly marked off than before, but those of the second and third pairs are ill-defined, their position being still marked only by pale crescentic areas. Extending the entire length of the germ band, from the stomodaeal invagination to the caudal end, is a wide shallow groove, the *neural groove*. This is bounded laterally by a pair of low ridges, the *neural ridges*. From the ectoderm included in the neural groove and the neural ridges the ventral nerve cord is formed.

The invaginations forming the silk glands have now lengthened out to tubular sacs extending caudad and slightly laterad, but as yet not reaching the posterior border of the first trunk segment.

The definitive number of tracheal invaginations on the trunk, ten pairs in all, are now present, and appear as narrow slit-like openings.

The stomodaeal depression has deepened, becoming funnel-form.

The invagination forming the anus, the *proctodaeum,* and those forming the Malpighian tubules, also appear at this stage, but are not apparent from the ventral side. Their formation will be more fully described in the sections devoted to the alimentary canal.

Stage X. Estimated at 54-56 hours (Figs. X and Xa). At this stage several well marked changes in the general appearance of the embryo are noticeable. The cerebral lobes do not now plainly show their subdivision into protocerebral and deutocerebral lobes, they form instead conspicuous hemispherical swellings, one on each side of the head, which on the dorsal side are marked off from the trunk by a deep constriction. The labrum (*Lm*) has become a conspicuous conical process projecting out from between the cerebral lobes parallel to the long axis of the embryo and is somewhat flattened dorsoventrally; its tip is seen to be plainly bilobed.

The stomodaeal invagination (*Sto*) has lengthened to a short tube, and its connection with the mid-intestine, whose outlines are now plainly visible, is much more apparent. The tritocerebral lobes are scarcely visible and are in fact in process of flattening out and disappearing. The antennal rudiments and mouth parts have changed but little, except that the latter have lengthened a little in a dorsal-ventral direction and are beginning to take on their characteristic forms, the mandibles being long papillate, while the two pairs of maxillae are shorter and more blunt, somewhat enlarged and flattened at the tip, like a pestle. The rudiments of the legs are now apparent as low rounded elevations in line with the mouth parts. The first pair are fairly well defined, the second and third pairs are faint, the last especially so. The germ band as a whole has increased considerably in width, now covering a trifle over a half of the entire surface of the yolk, and is divided by slight constrictions into thirteen segments, the definitive number.

The silk glands have lengthened to slender tubes extending caudad to the sixth trunk segment.

The openings of the tracheal invaginations, the future spiracles, are now small round aperatures and by the widening of the germ band have been carried so far laterad as to be virtually invisible from the ventral surface. The invaginations themselves have increased in size and extent and form flat sacs lying close against the inner surface of the ectoderm.

The proctodaeal invagination (Pro) has made its appearance as a funnel-shaped depression just dorsad of the caudal pole of the egg while the Malpighian tubules (Mal) are seen as two pairs of short blind tubes arising from its base.

Stage XI. Estimated at 56-58 hours (Fig. XI). The general appearance of this embryo is very similar to that of the stage preceding. The swellings representing the tritocerebrum have totally disappeared. The germ band covers a little over one-half of the circumference of the yolk. The thirteen segments of the trunk are more sharply separated from one another, and a lateral constriction slightly deeper than the rest marks the boundary between head and thorax. Both the silk glands and the Malpighian tubules have increased in length. The tracheal invaginations have each sent off tubular branches caudad, cephalad and ventrad. Those directed caudad and cephalad have united with the corresponding branches in the preceding and succeeding segments to form the longitudinal trunks, while the ventral branches are about to unite with those of the opposite side to form the ventral loops or commissures (TraCom). At this stage the latter are seen on the ventral surface as delicate sharply outlined tubes which as yet have not quite reached the mid-line.

Stage XII. Estimated at 60 to 62 hours (Fig. XII). The embryo now begins to resemble the larva decidedly in general

appearance. The head and trunk are sharply separated from
one another by a deep constriction. The head has become broader
and its sides rounder; the trunk segments are marked off from
one another by sharply defined constrictions. The germ band or
rather that part of it constituting the body wall of the future
larva has increased in width until it now virtually covers the entire
yolk, its lateral margins uniting along the dorsal mid-line. At the
same time the head has turned ventrad through an angle of about
forty-five degrees, so that the labrum, which formerly projected
nearly straight out in front, is now directed obliquely ventrad, and
the mouth, whose position up to this time corresponded to the
cephalic pole of the egg, is now brought to the ventral surface. A
similar change occurs at the posterior end of the embryo. The
proctodaeal invagination makes its first appearance on the dorsal
side of the posterior end of the egg (see Fig. Xa, *Pro*), the posterior end of the germ band since Stage VII having been curved
around the caudal pole of the egg, and the embryonic hind-intestine
up to this stage has accordingly had an oblique course, being directed obliquely caudad and dorsad. Now its direction corresponds
with the long axis of the embryo, and the anus having come to lie
precisely at the posterior end both of the embryo and of the egg.
Since at Stage VIII both ends of the germ band were seen to
have been bent around the two poles of the egg, these changes
therefore represent a partial straightening of the embryo, and
since the whole embryo does not appear to have materially increased in length, judging by its relation to the chorion, it must
be assumed that the ventral half of the embryo has contracted, at
least to a certain extent.

The labrum has meanwhile changed little in form. The tritocerebral swellings have completely vanished, the area formerly
occupied by them being now smoothly rounded. The antennae
are large and somewhat conical in form, and directed cephalad.
The mandibles have not perceptibly changed. The first maxillae,
however, have lengthened slightly and their peripheral portions are
plainly seen to be turned inward. The second maxillae have become bluntly conical in form, and slightly flattened dorsoventrally,
lying close to the ventral surface of the head, and directed cephalad. At the same time their bases have approached the ventral
mid-line, carrying with them the openings of the silk glands.

The three pairs of leg rudiments are plainly outlined and constitute low rounded protuberances. The ventral nerve cord is now
distinctly visible, on ventral view appearing as a dark band, segmentally enlarged, and pierced intersegmentally by oval openings.

The tracheal system is well on its way to completion, as indicated by the completion of the ventral tracheal loops (*TraCom*)
which are plainly seen on ventral view as delicate sharply outlined
tubes.

The silk glands and the Malpighian tubules, the latter no longer
visible from the ventral side, have continued to increase in length.

Stage XIII.   Estimated at 62-64 hours (Fig. XIII).   The
changes of most importance at this stage concern the antennae
and mouth parts. The former are less prominent than at Stage
XII, and are in fact in process of disappearing. Both the mandibles and first maxillae are bluntly conical in form and directed
inwards and forwards. The second maxillae have drawn still
closer together, and are becoming united to one another by their
mesial borders. At the same time the two external openings of
the silk glands are also brought close together. The second maxillae unite by their edges only, overarching, as it were, the intervening space on which the silk glands open, and which, after the
completion of the fusion of the second maxillae, becomes the
common unpaired duct of the silk glands which ultimately opens
at the tip of the labium.

Stage XIV.   66-68 hours (Fig. XIV).   At this stage the development within the egg is virtually completed. The principal
changes to be noted are: the almost total disappearance of the
antennal rudiments, of which only the faintest vestiges remain
visible on the exterior, and the completion of the union of the
second maxillae to form the labium. At 76-78 hours, eight hours
later—the embryo will rupture the chorion and elongate, being at
the same time flexed in a direction opposite to that of the embryo,
and become a young larva (Fig. XV) whose structure has already
been described.

3. *Segmentation*

It is generally believed by students of invertebrate morphology
that insects as well as other arthopods are made up of a linear
series of serially homologous segments, which in the primitive

ancestors of the highly specialized modern representatives of the arthropods were probably essentially similar to one another, much as they are in the modern chilopods or centipedes. Each of these component segments consists of an annular section of the body wall, provided with a pair of appendages, and containing a pair of ganglia and a pair of coelomic (mesodermal) sacs. In the insects, however, as well as in other modern representatives of the arthropods, a differentiation has taken place among these originally similar segments, accompanied also by a division of labor. In the insects, for example, a certain number of the anterior segments have become united to form the head, the appendages of some of these head segments having meanwhile become altered to form biting or chewing organs; other segments, such as those of the abdomen, although otherwise but slightly modified, have totally lost their appendages. For the most part the segments composing the insect body are easily recognizable, except in the cephalic region. Here the modification of the primative segments has been especially great, and the identification of these has for the past century been the subject of research, the problem having been attacked first from the anatomical side, later from the embryological. A complete account of the work done in this field cannot be attempted here, but may be found in such special papers as those of Heymons (1895a), Folsom (1900), Comstock and Kochi (1902), and Riley (1904).

In insect embryos the segments originally present are much more readily recognized than in the imago, the primative conditions being more or less imperfectly reproduced or recapitulated, and our present knowledge of the segmentation of the insect is principally based on embryological rearches. A hypothetical generalized insect embryo, combining the results of investigation on this problem is represented in figure 36.[16]

A primary head and a primary trunk division are recognized

[16] In this figure, borrowed from Snodgrass (1910), one segment of the abdomen is omitted, eleven only being represented. Heymons (1895) has shown that in the Dermaptera and Orthoptera twelve abdominal segments are present. The location of the mouth in the first segment may also be criticized, since it is the opinion of Comstock and Kochi (1902) that the mouth opening is located on the ventral surface of the third (tritocerebral) segment.

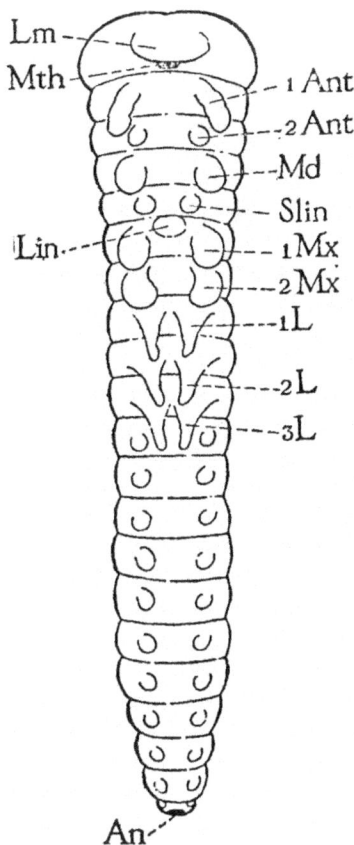

FIG. 36.   Diagram of a generalized insect embryo, showing the segmentation of the head, thoracic, and abdominal regions, and the segmental appendages.   From Snodgrass (1910).

in the insect embryo, as distinguished from the secondary or definitive head and trunk divisions of the imago.   The primary head division comprises those segments whose ganglia unite to form the supraoesophageal ganglion or brain.   The number of segments entering into this division has been variously estimated; Janet (1899), basing his conclusions on a study of the structure of the head of an adult wasp, *Vespa*, enumerates seven, while Carrière finds four in *Chalicodoma*, but it is probably safe to say that the majority of investigators follow Viallanes (1887), Patten (1889), and Wheeler (1893) in recognizing but three segments. These are: (1) the protocerebral, (2) the deutocerebral, and (3) the tritocerebral.   The first of these is without true appendages; the second bears the antennae, the third bears in some of the more primitive insects vestigal appendages probably corresponding to the second antennae of the Crustacea, as the researches of

Viallanes and others have shown. Following the three segments of the primary head division come three, or possibly four segments which enter into the composition of the head of the adult insect and which bear those appendages commonly known as the mouth parts (mandibles and maxillae). The next three segments, bearing legs, form the thorax. The remainder, twelve in number in primitive insects (Heymons, 1895a), of which only eleven are apparent in the majority of the embryos of pterygote insects, as in the honey bee, form the abdomen.

The early appearance of the definitive segments of the honey bee has already been described, but the precise identification of these is uncertain, even at Stage VII, since the segments of the primary head division are not clearly demarcated, and distinguishing characters, in the form of embryonic rudiments, are yet entirely lacking. At the Stage following (VIII), all of the segments, excepting the terminal segments of the abdomen, are readily distinguishable and may now be identified, since the rudiments of the appendages, and of the silk glands and spiracles have put in an appearance. The protocerebral and deutocerebral segments (1Br, 2Br) are plainly distinguishable, in fact more so than at any other time in the development. The protocerebral segment consists, as already described, of two large lobes extending half way down the sides of the egg, and rounded at their external margins. This segment possesses no recognizable appendages, although certain investigators have compared the labrum (Lm) which in many insects is more or less bilobed, with a pair of fused appendages. Heymons (1895), however, has pointed out that true appendages arise laterad of the nervous system, while the labrum in all cases arises in the mid-line between the two lateral halves of the nervous system; its homology with a pair of fused appendages therefore seems very improbable. The protocerebral segment, in addition to the labrum, bears the compound eyes and the ocelli, and contains a certain amount of mesoderm, but no coelomic sacs.

The deutocerebral segment (2Br) consists also of two large lobes. These are somewhat triangular in outline, and their external apices, which are, with respect to the germ band, directed laterad and somewhat caudad, bear the papillate antennal rudiments. In certain insects the antennae are post-oral in position at the time of first appearance, and in fact, in the honey bee a straight line drawn between them would intersect the germ band behind the mouth, yet their obvious connection with the deuto-cerebral lobes precludes their being seriously considered as any-thing but pre-oral in their relations in this insect. To this segment belong also well developed coelomic sacs, as will appear later.

The tritocerebral (second antennal, intercalary or premandibu-lar) segment ($3Br$) is, in the honey bee, of more than ordinary interest, but since Bütschli (1870) in the embryo of this insect first called attention to the presence of appendage-like swellings in front of the mandibles, between these and the antennae, thus occupying the region afterwards identified as that of the trito-cerebral segment. "Close in front of the anterior angles of the cephalic plates, between the mandibles and those angles, there arises a conical projection, springing from each of the germinal ridges, which attains a considerable degree of development and appears almost like a pair of inner antennae" (p. 538). Grassi (1884) also observed these appendages, considering them as the first pair of mouth parts. In another place (p. 201) he states "The first pair of mouth parts, which has an ephemeral existence, may possibly be compared to one pair of the antennae of the arthropods." Tichomiroff (1882) also noticed a pair of transitory swellings in front of the rudiments of the mandibles of the embryo of the silk worm. Wheeler (1893) de-scribed unmistakable vestigial appendages on the tritocerebral seg-ment of the collembolan *Anurida*. Heymons (1895), although expressing his disbelief in the presence of appendages on this segment in pterygote insects, recognized its presence in the Derm-aptera and Orthoptera and found that it was provided with rudi-mentary coelomic sacs in these orders. Later researches (Uzel 1897, Claypole 1898, Folsom 1900), have shown that true append-ages appear on the tritocerebral segment in several apterygote insects, and Uzel has found that in certain forms these appendages may even persist in the imago. Moreover Riley (1904) has de-scribed traces of appendages on the tritocerebral segment in the cockroach. In the honey bee the swellings on the tritocerebral segment first appear at Stage VIII, and attain their greatest development at the stage next following, Stage IX. During these stages they form conspicuous low conical swellings situated a trifle nearer the mid-line than the rudiments of the mouth parts, while a straight line connecting them would pass close to the posterior edge of the stomodaeal invagination. Examination of cross sections through the region of this segment at a stage a trifle older than Stage VIII demonstrates that the bases of these swellings are nearly in line with the bases of the rudiments of the mandibles and maxillae, but that these swellings do not resemble the rudiments of the mouth parts in their structure. If the appendage-like swellings on the tritocerebral segment at Stages VIII or IX are compared with the rudiments of the antennae or the mouth parts a difference in their histological character becomes at once apparent. The rudiments of the true appendages are ectodermal evaginations composed of numerous small and slender cells with a core of mesodermal tissue; those of the tritocerebral swellings on the other hand are composed of an outer layer of cells destined to form hypodermis, and an inner layer of neurogenic cells, continuous cephalad with the rudiments of the protocerebrum and deutocerebrum, caudad with those of the ventral cord. In brief, the tritocerebral swellings are composed principally of neurogenic tissue, and at a later stage the tritocerebrum is actually split off from the ectoderm of this region. With these facts in view the probability that these swellings represent appendages diminishes and it therefore seems more appropriate to consider them as exaggerated ganglionic swellings. At Stage X the tritocerebral swellings have become almost invisible, this region of the head having flattened out, and before Stage XI is reached the swellings have totally disappeared. Bütschli (1870) states that these rudiments fuse to form a sort of lower lip for the larva. An inspection of Stages VIII to X shows that this is not the case, but that there is on each side a fold which appears to be a continuation of the cephalic edge of the mandibular rudiments and which extends cephalad and mesiad and terminates on the external side of one of a pair of rounded eminences. An inspection of the transverse sections shows that these eminences are only the anterior ends of the neural ridges, as shown in the figures of these stages, and it seems also probable, judging from the relation of the parts, that these swellings belong rather to the mandibular than to the tritocerebral segment. Recently (1909) Hirschler has described in *Donacia* a pair of similar swellings, situated on the tritocerebral segment near the ventral mid-line,

which he terms "hypopharyngeal papillae," and which he believes
contribute to the formation of the definitive hypopharynx. These,
however, are clearly not to be considered as comparable to segmental appendages, for obvious reasons.

The three following segments, the first three of the primary
trunk division, the mandibular, and the first and second maxillary
segments, are in most respects similar to those of the embryos
of other pterygote insects. The appendages (*Md, 1Mx, 2Mx*)
up to the time of hatching are in the honey bee simple in
form, and do not exhibit any differentiation into lobes. Folsom
(1900) has discovered in *Anurida*, a primitive apterygote insect,
plain evidences of an additional segment intercalated between
the mandibular and first maxillary segment and bearing on its ventral surface a pair of papillae, representing a pair of rudimentary
appendages (Fig. 36, *Slin*). These fuse with a median tongue-
like evagination of the first maxillary segment (*Lin*) to form the
hypopharynx or lingua of the mature insect. This segment, with
its corresponding structures, is wanting in the embryo of the honey
bee as is apparently also the case in other pterygote insects.

The three segments next in order bear the legs (*1L, 2L, 3L*)
the rudiments of which appear on the surface, at Stage X, as low
protuberances, but are withdrawn from view shortly before
hatching.

The remaining segments, twelve in number, constitute the definitive abdomen of the adult in insects other than those of the
Hymenoptera. In this order, as is well known, the first abdominal
segment is usually united to the third thoracic segment (metatho-
rax) and may be said to constitute a part of the definitive thorax.
In the embryo of the honey bee, as in those of most of the higher
pterygote insects, the abdomen consists of eleven segments plus a
terminal or anal segment. Only ten abdominal segments anterior
to the terminal segment are usually visible from the exterior
(Figs. Xa and XI). Sections cut in the sagittal plane through
embryos of Stage XI-XII (Fig. 48A) show that just anterior to
the short terminal segment there is a short and much reduced
eleventh segment, from which a rudimentary pair of ganglia are
produced. The number of segments present in the abdomen of
the honey bee is therefore the same as that found by Heymons
(1895, 1895a) in the Dermaptera and Orthoptera.

The embryos of nearly all insects, as is well known, display at some time during development more or less evident rudiments of appendages on the ventral surface of the abdomen, as shown in figure 36. Those on the first segment of the abdomen are frequently very conspicuous and may, before the close of development, attain quite a considerable size, as in *Melolontha* (Graber 1890). Bütschli (1870) has described rudiments of appendages on all of the abdominal segments of the embryo of the honey bee, the last two being especially conspicuous. "The three thoracic segments are provided with short appendages projecting laterally and posteriorly, and by careful observation of the following trunk segments a similar but very ill-defined process may be observed on all of them; in the twelfth and thirteenth segments these appendages are so much developed that they can be designated as the anal appendages" (p. 537). Bütschli's statements however, are not well supported by his figures, since these do not show the abdominal appendages with sufficient definiteness to be convincing, with the exception of the last two, belonging to the ninth and tenth abdominal segments. These are quite definitely shown. Grassi (1884) apparently did not see the less well defined abdominal appendages described by his predecessor, but states that in several instances he saw two pairs of processes near the hinder end of the embryo, at a stage when the rudiments of the mouth parts had scarcely made their appearance. In Grassi's drawings (Tab. IV Fig. 10) these processes have much the appearance of artefacts, and his statement that these appendages frequently could not be seen, lends probability to this assumption. The writer was never successful in finding anything which could be safely construed as abdominal appendages. They certainly occur, nevertheless, in certain Hymenoptera. Carrière (1890) observed unmistakable rudimentary appendages on the first and second segments of the abdomen, and indications of them on the third and fourth segments. All of these soon vanish. The embryo of *Hylotoma* (Graber, 1890) possesses twelve pairs of well defined abdominal appendages, but this is to be expected, since *Hylotoma* is a much more generalized form than *Apis* or *Chalicodoma*. Both Wheeler (1910) and Tanquary (1913) report finding rudimentary appendages on the abdominal segments of ant embryos.

VIII

NERVOUS SYSTEM

1. *Newly hatched larva*

The nervous system of an insect consists of a *brain, ventral nerve cord,* and a complex of ganglia associated with the brain, the *stomatogastric system.* The ventral nerve cord and the brain of the young larva—Stage XV,— have already been briefly described but before entering upon an account of their development it will be desirable to examine their structure more closely. The ventral cord, as already said, consists of seventeen ganglia, one for each of the primary trunk segments. These ganglia are actually double, each representing a pair of simple ganglia united in the mid-line. The first three of these double ganglia, belonging to the mandibular and the two maxillary segments, are closely united to form a large elongated compound ganglion, the *suboesophageal ganglion* (Figs. XV, 38, 39, 42, 44 and 45, *SoeGng*), situated in the lower half of the head, below the oesophagus, and supplying nerves to the mouth parts. Its compound nature is clearly indicated by the presence of two transverse constrictions dividing the ganglion into three subequal parts (Fig. 39). Springing from the anterior end of the ganglion is a pair of stout connectives leading to the brain and embracing the oesophagus, the *circumoesophageal commissures,* while from its posterior end two similar connectives or commissures unite it to the ganglion next following, the first thoracic. The eleven ganglia following, including the first thoracic, are essentially alike except as to their size, the first thoracic being the largest, the second and third thoracic ganglia a trifle smaller, while the abdominal ganglia diminish slightly in size toward the posterior end. The last three ganglia are united to form a compound ganglion, but the last of the series, the eleventh abdominal, is so slight that it may be considered as vestigial. Each of the ganglia (Fig. 37, A, C and D) is flattened in form, about

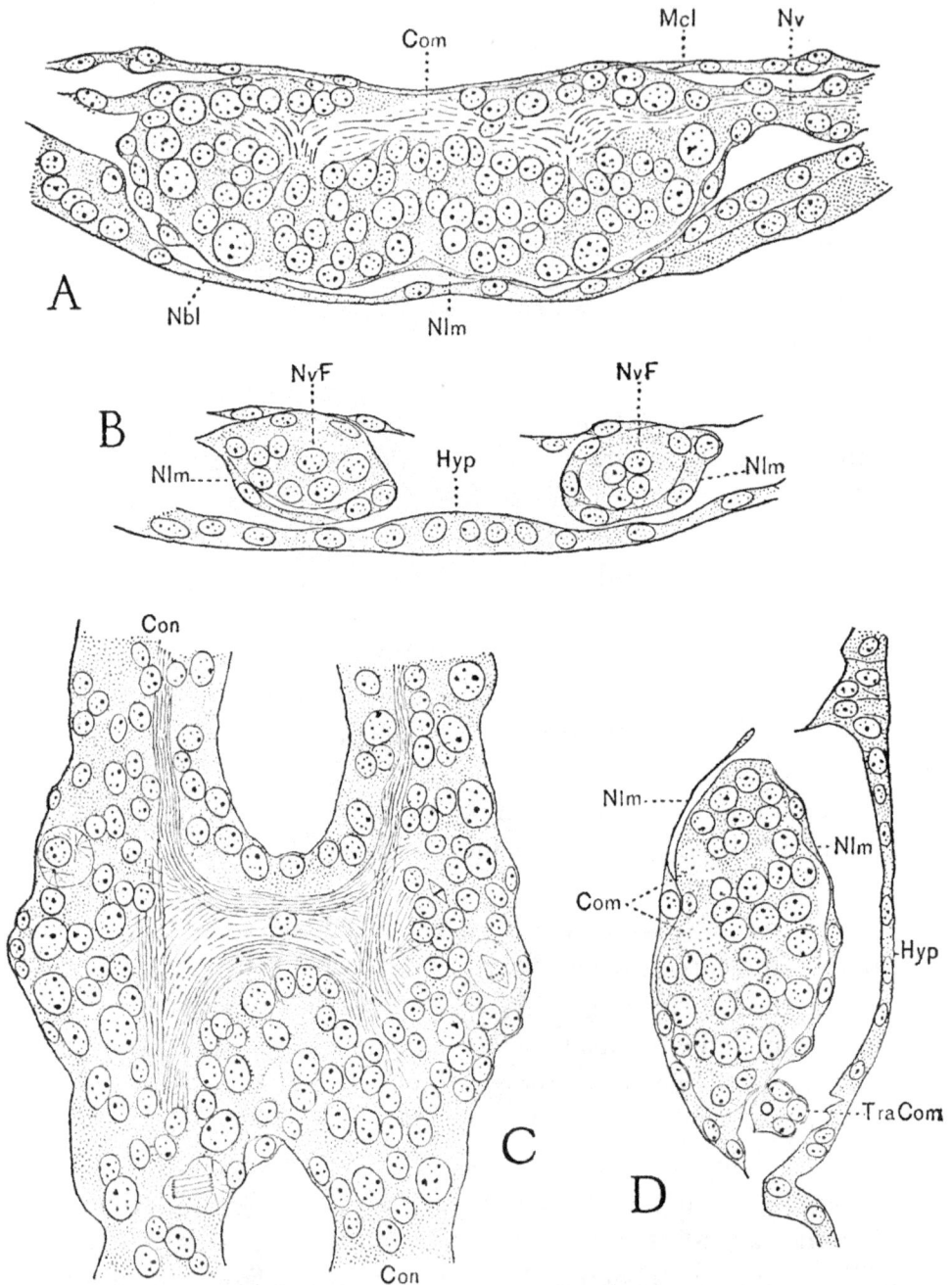

FIG. 37. Sections illustrating the structure of the ventral cord of a
newly hatched larva (Stage XV), x 900. A, transverse section through the
second thoracic ganglion, showing the lateral nerves (*Nv*), neurilemma
(*Nlm*), commissure (*Com*), and neuroblasts (*Nbl*). B, transverse section
through the connectives between the second and third thoracic ganglia,
showing nerve fibres (*NvF*) and neurilemma (*Nlm*). C, sagittal section
of the third thoracic ganglion. D, coronal (horizontal) section through
the third thoracic ganglion.

twice as broad as long, with a slight median depression on its ventral surface. From its anterior and posterior ends arise thick connectives joining it to the ganglia of the adjacent segments. Each ganglion also sends out laterally a large nerve trunk (Fig. 37A, *Nv*) which soon breaks up into smaller branches and these are distributed to various parts of the internal organs and to the body wall.

Histologically each ganglion is composed of a mass of cells, the *ganglion cells,* each of which sends off a slender *nerve fibre.* These fibres unite in the upper half of each ganglion with the nerve fibres from other ganglia to form bundles or strands (Fig. 37A). A large longitudinal strand traverses the lateral halves of the ganglia and passes through the center of the connectives; these two strands thus serve to link together all of the ganglia of the nervous system, including those of the brain (Fig. 37B and D). Two other smaller strands, the *transverse commissures* (Fig. 37A and D, *Com*) pass close together side by side from one side of each ganglion to the other thus joining its lateral halves. The nerve fibres in a ganglion consequently form a figure like the letter H, the cross bar of the H, however, being double (Fig. 37C). Each ganglion is thus incompletely divided into five regions; two situated laterad of the longitudinal fibres, and three in the mid-line. These Wheeler (1893) following Graber, has named the *lateral,* the *anterior,* the *central* and the *posterior gang-liomeres.* Other nerve fibres also pass out from each ganglion laterally, forming the core of the lateral nerves (Fig. 37A).

An exceedingly thin and delicate cellular layer, the *neurilemma,* (Fig. 37, *Nlm*) covers the exterior of the ventral cord and also of the brain.

The *brain* of an insect (Fig. 38) is to be regarded as composed essentially of three double (paired) ganglia: the *protocerebrum* (*1Br*), the *deutocerebrum* (*2Br*), and the *tritocerebrum* (*3Br*), corresponding respectively to the protocerebral, deutocerebral, and tritocerebral segments. These brain divisions consist of pairs of more or less evident swellings, each of which is made up of a mass of ganglion cells traversed by a thick central core of nerve fibres continuous with those of the longitudinal connectives of the ventral nerve cord. The protocerebrum of the young bee larva

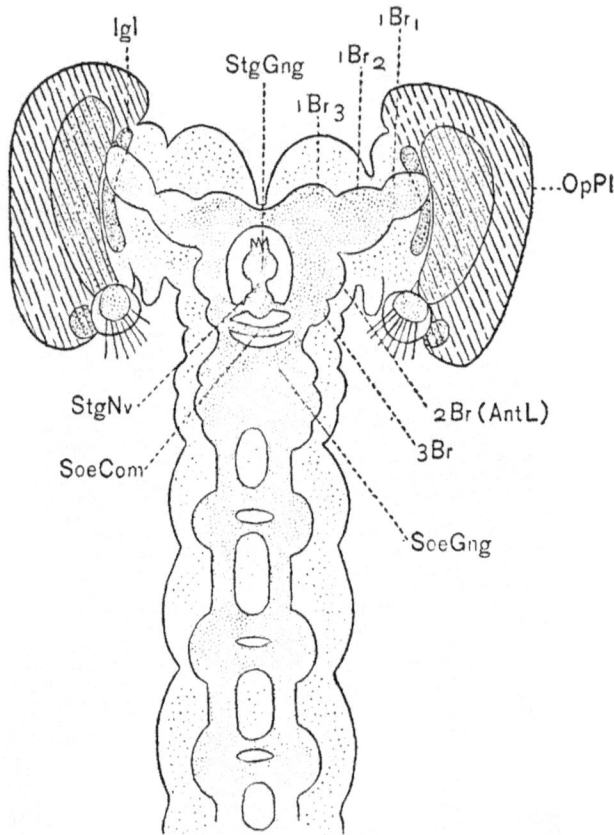

Fig. 38. Diagram of the nervous system of an embryo of *Mantis*, showing the head and first three trunk segments. From Viallanes.

includes the greater portion of the brain and is made up of a pair of large lobes (Figs. 39-44, *1Br*) situated in the upper part of the head capsule. These lobes are somewhat ovoid in form, their larger ends directed caudad and slightly laterad; their smaller anterior ends lie close to the anterior wall of the head. Each lobe is somewhat compressed at right angles to the external surface. At their anterior ends the lobes are closely apposed to each other and are here united by a thin bridge of nerve fibres, the *supraoesophageal commissure* (Figs. 43 and 44, *SupCom*). Probably some fibres from the second brain division, the deutocerebrum, also enter into this commissure.

Each half of the protocerebrum of insects is typically divided into three lobes, which in many cases, as in the Dermaptera and Orthoptera (Fig. 38) are marked off from one another by thickenings of the ectoderm forming the wall of the head capsule.

Moreover, as well be seen by consulting the diagram represented in figure 38 the halves of the protocerebrum are typically directed laterad as well as cephalad.   In the larva of the bee the halves of the protocerebrum are, as mentioned above, directed caudad and laterad (Fig. 55).   This difference is due to the bending of the anterior end of the germ band and of the resulting embryo around the cephalic pole of the egg and must be constantly kept in mind in comparing the development of the brain of the bee with that of other insects.   Those forms in which the development of the brain has been most thoroughly worked out belong princi- pally to the order Orthoptera, and in comparing the embryonic development of the brain of the bee with that of those insects, another important point of difference must be also considered. In the Orthoptera, for example, the development is without the intervention of a larval stage, that is, direct from the embryo to a form completely equipped for active life, and provided, like the mature insect, with functional eyes, antennae and appendages. In the bee, on the other hand, the embryonic development termin- ates in a larva, a form not adapted to an independent existence, and very different structurally from the imago.   Many parts of the larva are correspondingly but slightly developed, and may be said to linger in a more or less latent or embryonic condition until near the time of pupation.   This is especially true of the brain.   At the time of hatching, the brain of the bee is, generally speaking, only comparable to the brain of the embryo of those insects with a direct development.

The three lobes of the protocerebrum of the insect brain have the following fate, as determined by Viallanes (1891), Wheeler (1893) and Heymons (1895) for the Orthoptera and Dermap- tera.   The first (outermost or anterior) forms the optic lobe, the second the optic tract, while the third lobe forms those parts of the imaginal brain included in the *protocerebral lobes*.   On account of its relatively undeveloped condition and the absence of ectodermal thickenings these three divisions of the protocerebrum of the bee are not marked off with the same clearness as in the Orthoptera and Dermaptera, nevertheless three lobes are distin- guishable, both on surface view (Fig. 39) and in longitudinal section (Figs. 43 and 44).   The first lobe ($1Br_1$), morphologically

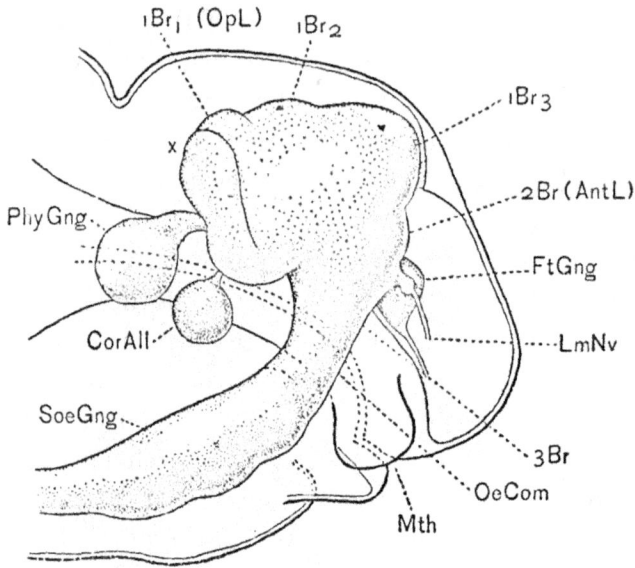

FIG. 39. Cephalic portion of the nervous system of a newly hatched larva (Stage XV), lateral aspect, drawn with the camera lucida from an entire preparation, and corrected with the aid of sections, x 243.

the anteriormost (Fig 38), but in the bee the posteriormost, is the broadest of the three, and subspherical in form. Its posterior half, separated from the rest of the brain by a groove, is com-

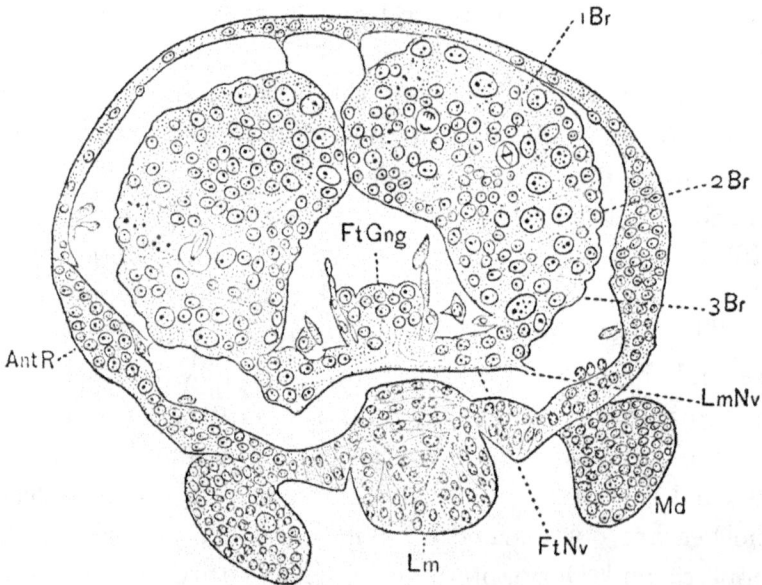

FIG. 40. Transverse section through the head of a newly hatched embryo (Stage XV), passing just in front of the mouth, showing part of the protocerebrum (*1Br*) and the frontal ganglion (*FtGng*), x 337.

posed of a thick plate of cells whose cytoplasm in many preparations stains more deeply than the remainder of the brain, and
whose nuclei are unusually large, clear, and sharply stained (Figs.
39 and 44 x).

The second or middle lobe of the protocerebrum ($1Br_2$) is
very short, its length being about one-half its width.

The third lobe ($1Br_3$) is about one-third the length of the corresponding lateral half of the protocerebrum and at its anterior
end meets its mate of the opposite side and is here united to it
by the supraoesophagael commissure (Fig. 43, SupCom).

The *deutocerebrum*, or *antennal lobes*, the second division of
the brain (Figs. 39, 40 and 41, $2Br$ [AntL] manifests itself externally as a pair of conspicuous rounded swellings of the antero-
lateral face of the brain just above the oesophagus and below
the protocerebral lobes. From the fibrous core of each of the

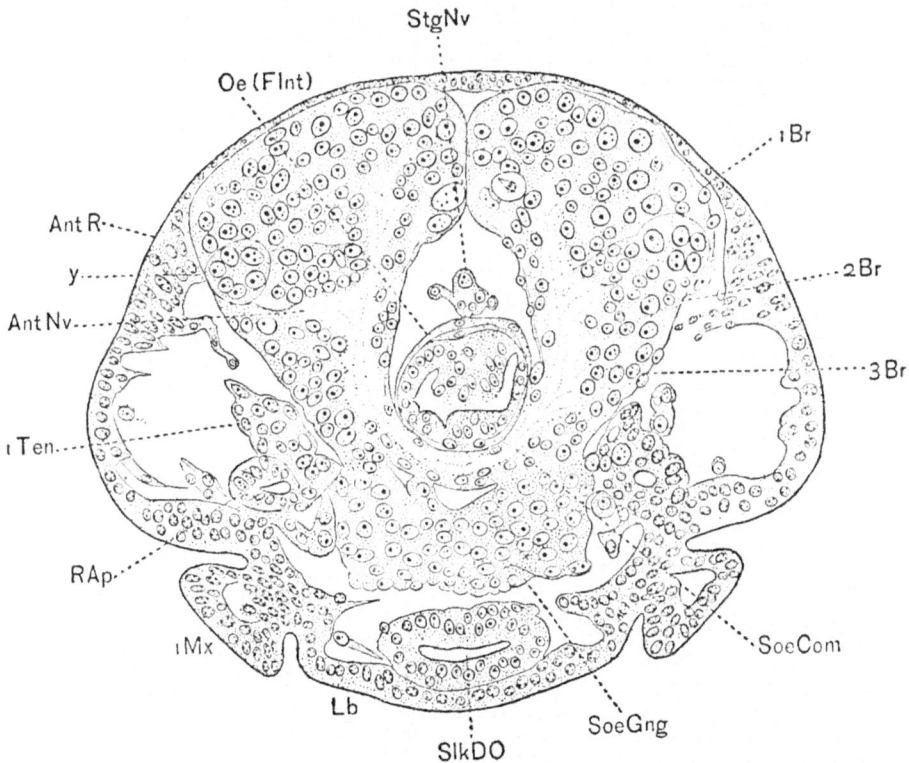

Fig. 41. Transverse section through the head of a newly hatched larva
(Stage XV) intersecting the circumoesophageal commissure and showing
part of the protocerebrum ($1Br$), the deutocerebrum ($2Br$), the tritocerebrum ($3Br$), and the anterior end of the suboesophageal ganglion
(SoeGng), x 337.

deutocerebral lobes is given off a thin strand of nerve fibres, the
rudiment of the antennal nerve (Fig. 41, *AntNv*) which passes
laterad to the antennal rudiment (*AntR*).

The *tritocerebrum* is in the larva, as in the adult, a rather ill-
defined region of the brain; its two halves are virtually continuous
ventrad with the halves of the circumoesophageal commissures,
dorsad they are fused with the antennal lobes or deutocerebrum.
Each is united to its mate of the opposite side by a long and thin
band of nerve fibres, the *suboesophageal commissure* (Figs. 41

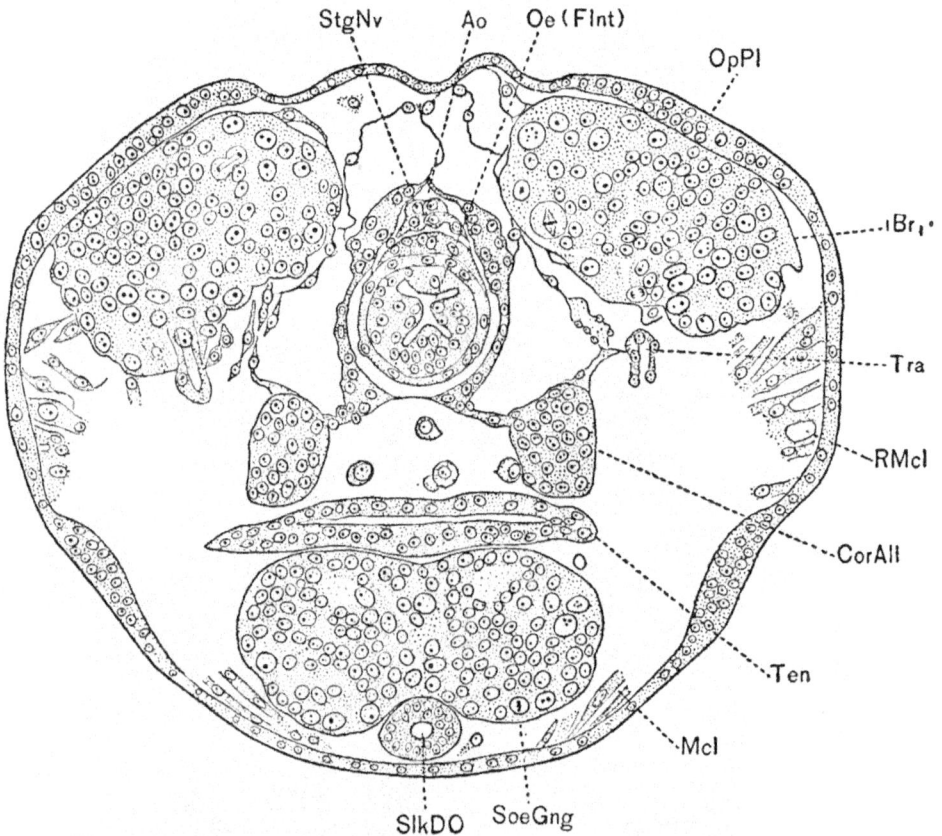

Fig. 42. Transverse section through the posterior region of the head of
a newly hatched embryo (Stage XV), intersecting the optic lobes (*1Br,
OpL*), the suboesophageal ganglion (*SoeGng*), and the corpora allata
(*CorAll*), x 337.

and 45, *SoeCom*), which passes beneath the oesophagus and intersects the median plane between the oesophagus and the anterior
end of the suboesophageal ganglion. The two halves of the tritocerebrum are in addition connected with one another and with

the *frontal ganglion* (Figs. 39 and 40, *FtGng*) lying between them, by short thick cords of nerve fibres covered on the exterior by a layer of ganglion cells, the *frontal nerves* (Fig. 40, *FtNv*). The two nerves composing this commissure arise on each side from the anterior face of the tritocerebrum and at their point of origin from the latter is the root of a nerve which passes cephalad to the labrum, the *labral nerve* (Figs. 39 and 40 *LmNv*).

The histological structure of the brain is simple and essentially similar to that of the ganglia. Each lateral half consists of a mass of ganglion cells covered by the neurolemma and surrounding a central mass of nerve fibres continuous with that of the circumoesophageal commissures.

The *stomatogastric system* of the newly hatched larva consists of the following parts: the *frontal ganglion*, the *stomatogastric*

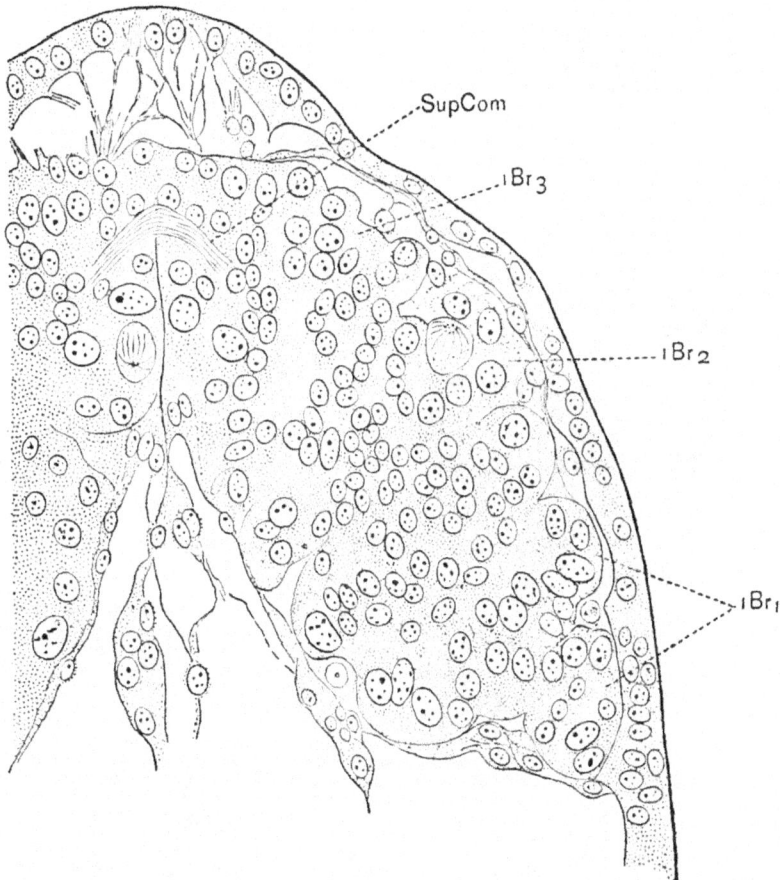

FIG. 43. Coronal (horizontal) section through the right half of the protocerebrum of a newly hatched larva (Stage XV), x 337.

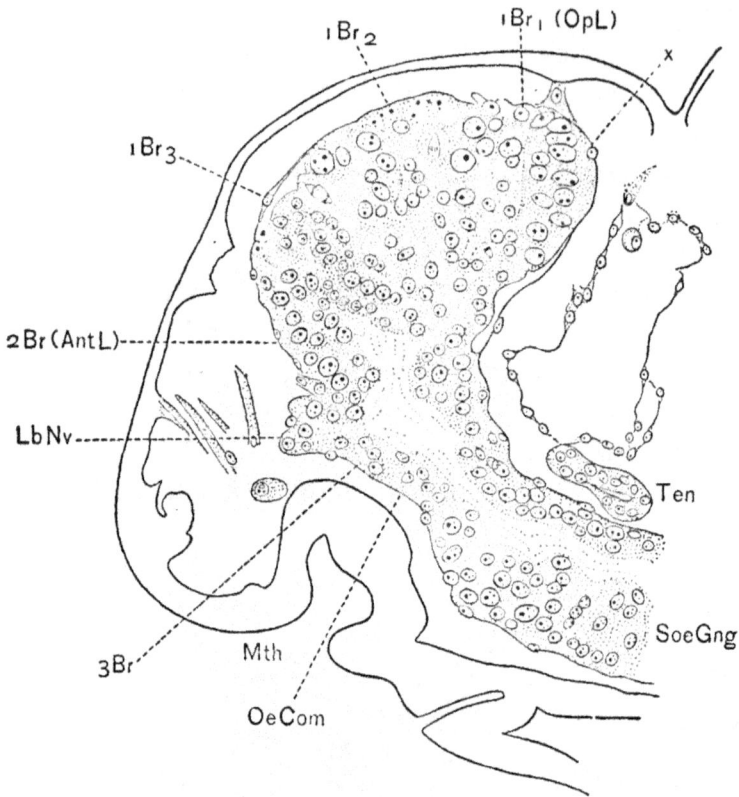

Fig. 44. Sagittal section of the head of a newly hatched embryo (Stage XV), passing laterad of the median plane and intersecting the right half of the brain (*Br*), the circumoesophageal commissure (*OeCom*) and the suboesophageal ganglion (*SoeGng*), x 337.

nerve or nervus recurrens, and a pair of ganglion lying against the sides of the posterior region of the oesophagus, the *pharyngeal ganglia*. To these may be added, for the sake of convenience, the so-called *ganglia* or *corpora allata* of Heymons, which in fact are probably not ganglia at all, according to Heymon's own statement (1897), (Figs. 39 and 42, *CorAll*).

The *frontal ganglion* (Figs. 39, 40 and 45, *FtGng*) is a large and conspicuous pyriform ganglion with its blunt end directed cephalad, lying in the median plane at the base of the labrum and just above the oesophagus. It consists of a compact mass of ganglion cells surrounding a core of nerve fibres. From its anterior end nerve fibres, accompanied by a few scattered ganglion cells, run to the tip of the labrum. At its posterior end the frontal ganglion becomes continuous with the *stomatogastric* or *recurrent nerve* (Figs. 41, 42 and 45, *StgNv*) which runs backward in the median

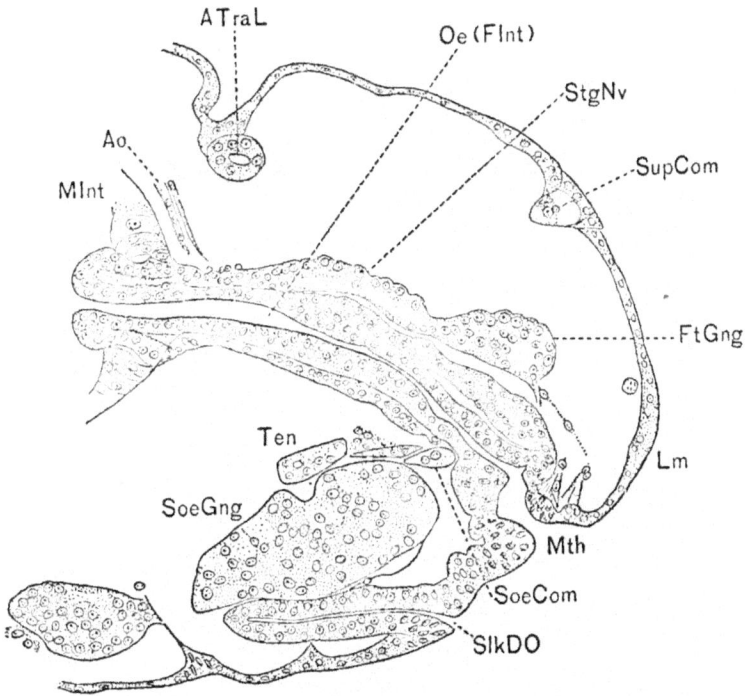

FIG. 45. Median sagittal section through the head of a newly hatched larva (Stage XV), intersecting the supraoesophageal commissure (SupCom), the frontal ganglion (FtGng), the stomatogastric nerve (StgNv), the oesophagus (Oes), and the suboesophageal ganglion (SoeGng), x 243.

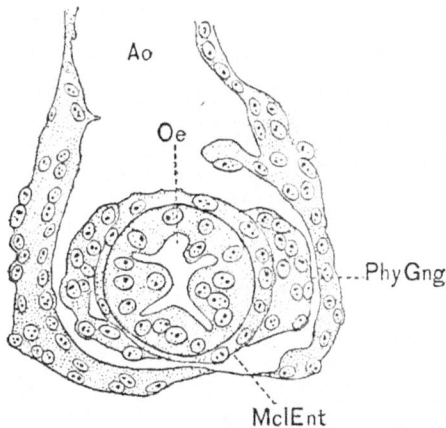

FIG. 46. Transverse section of the oesophagus, showing the pharyngeal ganglia (PhyGng), x 534.

plane between the two halves of the brain just above the oesophagus. This nerve consists, near its point of origin, of a thick bundle of nerve fibres surrounded by a sheath of ganglion cells, but diminishes rapidly in calibre as it passes caudad. At a point slightly cephalad of the junction of the oesophagus and the mid-intestine the recurrent nerve divides to form two small and relatively insignificant ganglia, the *pharyngeal ganglia* (Figs. 39 and 46, *PhyGng*). These are flat and lie closely applied to the lateral faces of the oesophagus.

The *corpora allata* (Figs. 39, and 42, *CorAll*) are two conspicuous spherical masses of cells which most resemble ganglion cells, but lack nerve fibres. These bodies lie some distance apart in the posterior part of the cavity of the head, just below the level of the lower edge of the oesophagus. These bodies are attached dorsally to the thin walled sacs which constitute the coelomic sacs of the antennal segment.

2. *Development*

Before entering upon a description of the development of the nervous system of the honey bee it will be profitable and in fact almost necessary first briefly to describe the course of development of the nervous system in the Orthoptera, a relatively primitive group, in which the intimate details of the process reveal themselves sharply and distinctly. The broader features of the development of the nervous system in the honey bee are simple and readily understood; the minuter details of the origin of the development of the nervous system, in other words its histogenesis, is on the other hand much more difficult to follow and relatively obscure, in fact so much so that it would be decidedly difficult of interpretation without the assistance afforded by previous work on this subject.

The nervous system of the insect arises, as was first shown by Hatschek (1877), from two longitudinal thickenings or ridges extending along the ventral side of the germ band, one on each side of the mid-line. These ridges, the *neural ridges* (Fig. 47, *NlR*) enclose betwen them a narrow median groove, the *neural groove* (*NlG*), the internal infolded portion of which forms the *median cord*. From the intrasegmental portion of the median cord and the neural ridges the ganglia are formed by a process of splitting off or delamination, the outer portion of the ectoderm remaining as the ventral hypodermis; in the interganglionic re-

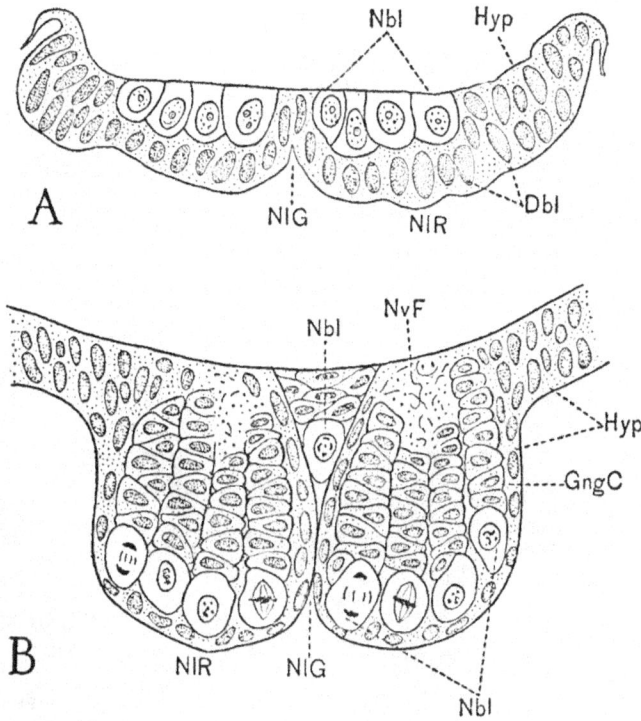

FIG. 47. Transverse sections through the ventral ectoderm of embryos of the grasshopper, *Xiphidium*. A, early stage, showing segregation of the neuroblasts (*Nbl*); B, late stage, showing the neuroblasts and their products, the ganglion cells (*GngC*). From Wheeler (1893).

gions the neural groove remains intact as hypodermis, the neural ridges here forming the connectives.

The histogenesis of the neurogenic tissue constituting the neural ridges and neural groove has been studied by several investigators, but Wheeler (1891, 1893) was the first to give a complete and consistent account of this process, as observed in the grasshopper *Xiphidium*. The essential features of Wheeler's results have since been confirmed by a number of investigators of the Orthoptera and also of other orders, among these are Heymons (1895) for the Dermaptera and Orthoptera, Carrière and Bürger (1897) for the mason bee, and Lecaillon (1898) for the beetle *Clytra*. Wheeler's account is briefly as follows:

At a stage soon after the completion of the formation of the germ layers the ventral ectoderm of the embryo is seen to consist of two different kinds of cells; a small number of large pale cells with vesicular nuclei, and a large number of smaller cells with small dark nuclei. The former will give rise to the nervous system and are therefore termed *neuroblasts* (Fig. 47A, *Nbl*), the latter form only the hypodermis and the structures associated with it; these cells as therefore termed *dermatoblasts* (*Dbl*). A little later, when the neural ridges and neural groove make their appearance, the neuroblasts become arranged in four or five rows on either side of the mid-line, lying next to the inner surface of the ectoderm. The neuroblasts next begin to bud off by mitotic division smaller and darker cells from their inner surface. Fig. 47B). These constitute the *ganglion cells* (*GngC*). Each neuroblast thus gives off by division a number of ganglion cells which together form a colum placed at right angles to the outer surface of the embryo, each column having at its base the parent neuroblast. Besides these neuroblasts lying on either side of the mid-line there is another set, the *median cord neuroblasts,* which are situated in the mid-line, at the bottom of the neural groove (Fig. 47B.) These neuroblasts are intersegmental in their arrangement, one neuroblast being placed between every two segments. Each median cord neuroblast gives rise to a heap of ganglion cells lying on its inner surface. These latter become displaced cephalad and form the posterior median portion of each ganglion. The remainder of the ganglion is formed from the invaginated ectoderm of the neural groove. From this portion, Wheeler believes, arise the cells which form the inner and outer neurolemma. The ganglion cells are the functional nerve cells of the ganglia and send off delicate processes which branch and thus form the nerve fibres or fibrillar substance which constitutes the connectives and the commissures.

The two divisions of the nervous system, the brain and the ventral nerve cord form a continuous whole from their inception, but for convenience they will be treated separately, following the usual custom. The development of the brain is slightly in advance of that of the ventral cord and, in the latter, following the general rule among arthropods, development progresses from the cephalic to the caudal end. In the bee, however, the development of the ventral cord is almost simultaneous in all of the segments, the anterior being but slightly in advance of the posterior.

A. The Ventral Cord

At Stage VII, when the formation of the germ layers is virtually completed the ectoderm of the germ band caudad of the procephalic lobes forms a uniform and rather thick layer composed of prismatic cells and covers the ventral half of the egg (Figs. 32, 48A). The cells of the ectoderm are not, however, all alike, those comprising its middle third having nuclei considerably larger and somewhat clearer than those of its lateral portions; this median area may be termed the *neurogenic area*, since it includes the cells from which the ventrad cord will arise. In the mid-ventral line there is a narrow strip, three to five cells wide, which is quite well defined in the anterior region of the germ band, less so in the posterior regions, the *median cord* (Fig. 48A, *MC*). With this exception the cells comprised within the neurogenic area are at this stage precisely similar as regards form, size, or staining reaction.

At the next Stage, VIII (Figs. 46B and C) several notable changes are evident. The ectoderm is greatly reduced in thickness laterad of the neurogenic area and also along the ventral mid-line. The neurogenic area therefore comprises two longitudinal thickenings, the *primitive swellings* (Fig. 48B, *PriSw*) which are separated by a median furrow, the *neural groove* (*NlG.*) The primitive swellings do not owe their formation to cell proliferation, but to changes in the form of the cells of different areas, the primitive swellings alone preserving the thickness of the ectoderm and the primatic form of its cells existing at Stage VII.

The primitive swellings, also shown in figures VIII, IX and X, are highest in the gnathal region, and diminish rapidly in height caudad, becoming scarcely distinguishable in the abdominal region. Moreover they are broader and slightly higher in the intrasegmental than in the intersegmental regions, as a comparison of figures 48B and C, and 48D and 49A will show, being thus divided into a series of segmental ganglionic swellings corresponding to the future ganglia.

The median cord can now be readily identified; its cells form the floor of the neural groove. At this stage the median cord is much broader and thinner than either at the stage preceding, or

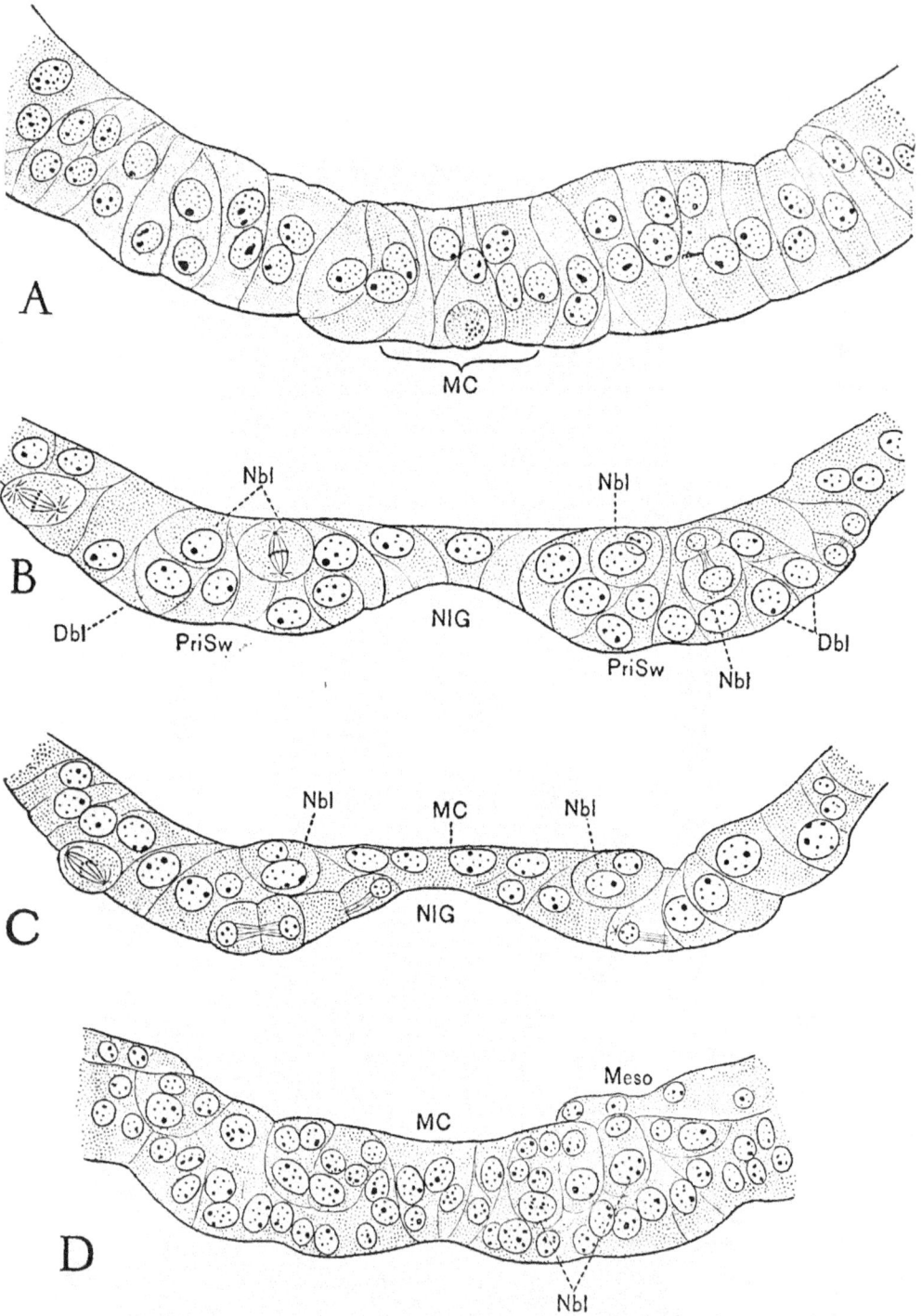

FIG. 48. Transverse sections through the ventral ectoderm of embryos of the honey bee, showing segregation of the neuroblasts (*Nbl*) and the dermatoblasts (*Dbl*), and the formation of the ganglion cells (*GngC*) by the granddaughter cells of the neuroblasts. A, thoracic region, Stage VII; B, middle of second thoracic segment, Stage VIII; C, anterior half of second thoracic segment, Stage VIII; D, second thoracic segment, Stage IX.

at subsequent stages, and has the appearance of being stretched laterally. This is especially marked in the preparation from which figures 48B and C were drawn.

The cells of the primitive swellings only in part retain the prismatic form seen at Stage VII; many of them are now seen to be either in the act of division or to have recently divided. These cells, fall into two classes, according to their position or to their mode of division. First there are those in which the division is equal, the mitotic spindle lying near and parallel to the external surface and the resulting daughter cells occupying a superficial position. These cells (*Dbl*) correspond to the "*dermatoblasts*" of Wheeler and are destined to form hypodermis only. Second, there are cells in which the division is unequal, the spindle lying near the internal surface and usually directed at right angles to it, the products of division consisting of a smaller central and a larger peripheral cell. The larger cells in these pairs correspond to the "*neuroblasts*" (Fig. 48B and C, *Nbl*) the smaller to the "*ganglion cells*" of Wheeler. A difference in the arrangement of the products of the dermatoblasts and neuroblasts becomes at once evident; those of the dermablasts form ordinary epithelial cells lying side by side; on the other hand each neuroblast together with its daughter cells form a compact and more or less ovoid nest or mass of cells (Fig. 48B and C).

At Stage IX, (Fig. 48D and 49A) the neural groove is both narrower and shallower than at the preceding stage, while at the same time the median cord has become correspondingly narrower and deeper, its cells resuming their characteristic long prismatic form. In cross section the median cord now presents somewhat the form of the letter V, its outer end forming the point, although in fact it is more or less truncate or flattened and constitutes the floor of the neural groove. The cells of the median cord are therefore long tapering in form, their inner ends always being wider than their outer.

At this stage the neuroblasts and dermatoblasts have become completely segregated from one another, so that each primitive swelling is divided into an inner neurogenic layer composed of cell nests, the *lateral cord*, made up of the neuroblasts and the ganglion cells, and an outer dermatogenic layer made up of the products of the dermatoblasts. The latter constitutes a single tier of cells which are growing progressively smaller by division, as evidenced by the diminishing size of their nuclei. The form of these cells ranges from cubical, near the mid-line, to long prismatic at the lateral edges of the primitive swellings.

The neuroblasts, which are frequently to be observed in division (Fig. 49A) are true teloblasts and suffer no diminution in size and are becoming increasingly conspicuous owing to the reduction in size of the dermatogenic cells. The outermost row of neuroblasts of each side is now seen to differ from the remainder in giving off its daughter cells mesiad instead of centrad. In the honey bee the arrangement of the neuroblasts is not quite so regular as in many other insects, the Orthoptera for example, but as a rule there are from three to five rows of neuroblasts in each lateral cord. The number of rows is greatest in the middle of each segment, as may be seen by comparing figures 48D and 49A. This difference exists in *Forficula* (Heymons 1895) and *Donacia* (Hirschler 1909), and is quite probably of frequent occurrence.

The number of ganglion cells has now increased, but not uniformly, since some cell nests possess as many as four, others only one.

At Stage X (Figs. 49B, C, and D) several changes are noticeable. One of the most evident is the flattening of the ventral surface of the embryo, resulting in the almost complete disappearance of the external evidences of both the primitive swellings and neural groove except in the gnathal region (Fig. X). As a result of the flattening of the external surface of the ventral ectoderm the lateral cords are, so to speak, thrust inward and project into the body cavity, especially in the intrasegmental regions. This change in the contour of the ventral ectoderm is probably brought about by the absorption of yolk and the consequent diminution of its mass, causing its withdrawal from the ventral ectoderm and a lowering of pressure upon the latter from within. The mesial edges of the two halves of the mesoderm now extend up to the lateral boundaries of the lateral cords and about as closely against them (Fig. 49C) that it is frequently not easy to distinguish the neurogenic from the mesodermal tissue,

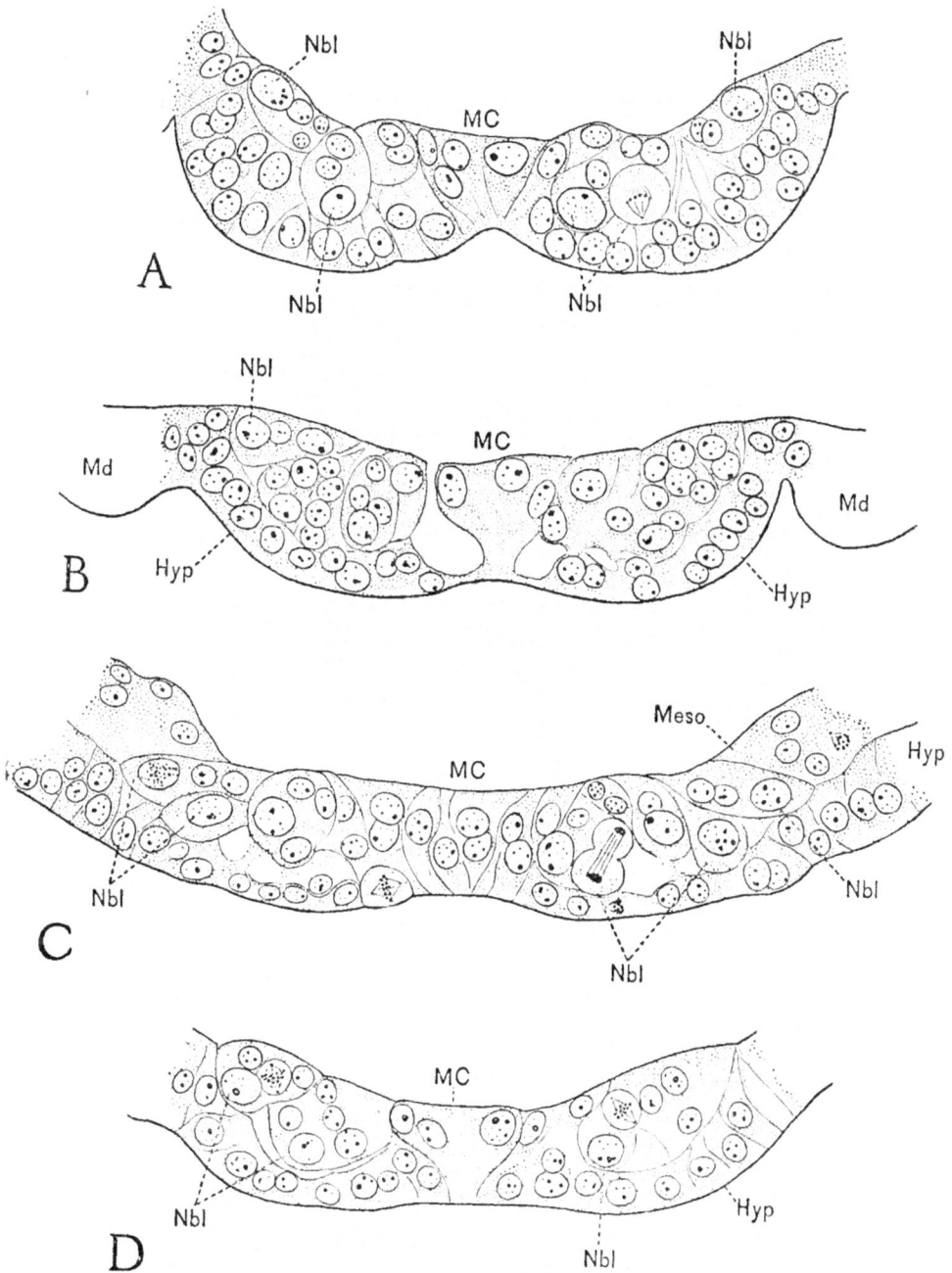

FIG. 49. Transverse sections through developing ventral cord. A, region between the first and second thoracic segments, Stage IX; B, first mandibular segment, Stage X; C, first thoracic segment, Stage X; D, posterior region of second thoracic segment, Stage X, x 600.

especially since the ganglion cells are very similar to those of the mesoderm.

The separation between the lateral cords and the hypodermal (dermatogenic) layer is more evident than before, being marked in many sections by irregular clefts (Fig. 49C). The cells of the hypodermal layer are still diminishing in size and now begin to form an evident and well defined epithelium. The cell nests of the lateral cords on the other hand, are becoming compacted together to form the definitive ganglia. This is particularly noticeable in the gnathal region (Fig. 50A) and at the anterior and posterior ends of the thoracic segments (Fig. 49D).

In addition to the mitotic figures of the neuroblasts it is not unusual at this stage to find other mitotic figures within the cell nests of the lateral cords. Two of these mitotic figures are to be seen in the section illustrated by figure 49D. Both the size of these mitotic spindles and their relation to the adjacent cells make it clear that they belong to the daughter cells of the neuroblasts. It is not clear whether all of the daughter cells thus undergo mitotic division but it is at least certain that many of them do. Wheeler, in his earlier paper on the histogenesis of the nervous system of the grasshopper (1891), believed that the daughter cells of the neuroblasts did undergo division. This opinion was relinquished in his later paper (1893) on the same subject, as regards the grasshopper, but not as regards *Doryphora*, in which form, Wheeler expressly states, the daughter cells of the neuroblasts divide. On the other hand Heymons (1895) for *Forficula* and Orthoptera, Lecaillon (1898) and Hirschler (1909), for the chrysomelid beetles, and Escherisch (1902) for *Musca*, deny the division of the neuroblast daughter cells.

At the stage next illustrated, Stage XI-XII, the entire ventral cord—with the exception of its extreme posterior end—becomes completely severed from the ectoderm and the ganglia begin to assume their final shape. The separation of the lateral cords from the overlying hypodermis was already evident at Stage X, but at that stage the median cord was still embedded in the hypodermis, the distal ends of the component cells of the cord forming the bottom of the neural groove. These cells have meanwhile been increasing slowly by mitotic division. These divisions are equal, no evidences of median cord neuroblasts having been observed. Meanwhile the median cord has also separated from

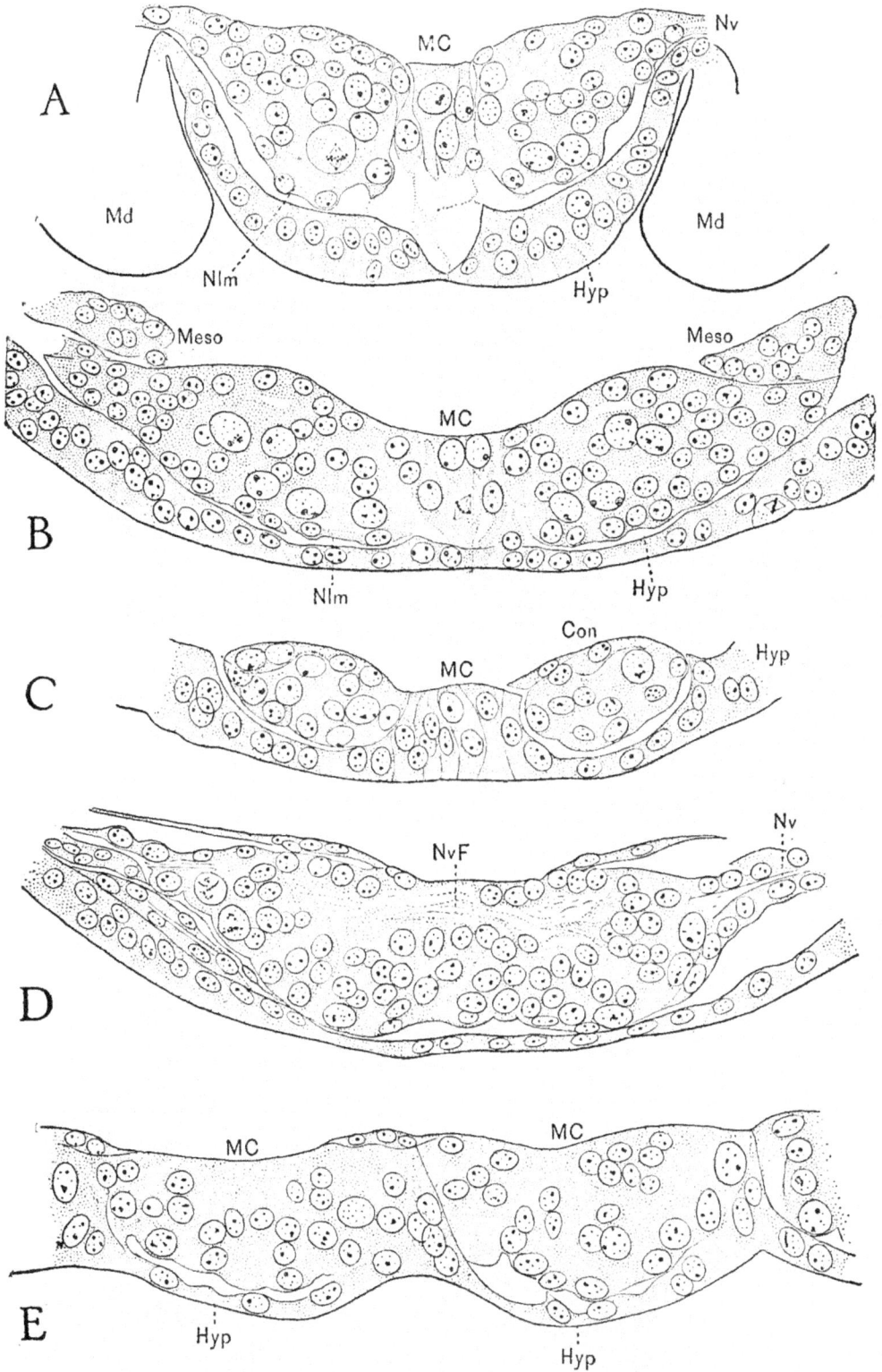

FIG. 50. Sections through the embryonic ventral cord, x 600. A, gang-lion and lateral nerves of the mandibular segment, Stage XI; B, ganglion of second thoracic segment, Stage XI; C, connectives between the second and third thoracic ganglia, Stage XI-XII; D, ganglion and lateral nerves of second thoracic segment, Stage XIV; E, median sagittal section through the ganglia of the first and second thoracic segments.

the hypodermis in the intrasegmental regions and now constitutes the median portion of each ganglion. This separation of the median cord from the hypodermis takes place by a progressive attenuation of the already narrow outer ends of the cells of the median cord until the external portion of the latter becomes reduced to thin strands, the hypodermis meanwhile closing in from both sides until its lateral halves meet and unite in the mid-line. The hypodermis along the ventral mid-line therefore owes its origin—in the intrasegmental regions, at least,—to hypodermal cells lying originally laterad of this region, as first noted by Grassi (1884). Similar conditions also exist in many other insect embryos, *Melolontha* (Graber 1890) for example. The median cord, after its separation from the hypodermis moves inward, frequently leaving behind it a temporary median notch or groove on the inner surface of the hypodermis, as shown in Fig. 50A. In this figure also are to be seen the delicate strands of protoplasm connecting the median cord and the hypodermis. This separation does not occur simultaneously throughout the length of each segment, but appears to take place first in its posterior half (Fig. 50E). In the intersegmental regions the median cord remains an integral part of the hypodermis (Fig. 50C, (*MC*); the lateral cords here separate from both the median cord and the hypodermis proper and constitute the connectives *Con*).

In the lateral cords the number of ganglion cells has greatly increased. In every segment a conical group of these lies along each side of the lateral cord, laterad of the neuroblasts, while a well defined single layer covers them on the ventral side. The lateral cell groups apparently furnish the material for the formation of the lateral nerves, while the ventral layer provides the material of the neurilemma. In the dorsal region of the lateral cords a cleft is apparent, separating the uppermost tier of cells from the remainder. These clefts are, so to speak, the forerunners of the nerve fibres, which during the succeeding stages traverse these spaces. Laterad of each ganglionic rudiment the mesoderm (Fig. 50B, *Meso*) presses close against it in the middle region of each segment, frequently, as at Stage X, making it difficult to accurately determine the lateral limits of the ganglion cells.

The final changes leading to the functional larval ganglion are illustrated by figure 50D, Stage XIV. The most important changes are: the development of nerve fibres, the development of the lateral nerves, and the final incorporation into the ganglia of the intraganglionic portions of the median cord.

The development of the nerve fibres corresponds essentially to the accounts of Wheeler (1893) and Heymons (1895). The ganglion cells of both the median and the lateral cords at about Stage XII become pyriform; the smaller pointed ends of the ganglion cells of the lateral cords being directed toward the clefts already described, while the smaller ends of the cells of the median cord are directed more or less dorsad. The smaller ends of the ganglion cells now become more and more attenuated and finally elongate into delicate protoplasmic fibres, the nerve fibres. During this process, in those regions where the commissures are formed, a vacant space also appears in the dorsal half of the median cord, in line with the spaces in the lateral cords, with which it unites. These spaces are traversed from side to side by the cell processes which are to constitute the nerve fibres of the transverse commissures (Fig. 50D, NvF). These commissural nerve fibres appear to arise from both the cells of the median and lateral cords, but mainly from pyramidal groups of cells situated immediately laterad of the median cord (see also Fig. 37A).

The lateral nerves are apparently formed from the aggregations of ganglion cells which at Stage XI-XII (Fig. 50B) constitute the extreme lateral portions of the ganglionic rudiments. These lateral cell groups are presumably derived by migration of ganglion cells which formerly lay within the semicircle formed on each side by the neuroblasts, since there is no evidence that cells are ever budded off from the neuroblasts in a laterad direction. Moreover both the arrangement of these lateral cells, and the fact that the number of ganglion cells lying immediately dorsad to the neuroblasts is perceptibly diminished when the lateral cell groups appear, tend to confirm this conclusion. The fate of the lateral cell masses is clearest in the gnathal segments, since in these the lateral nerves are short and straight, so that the entire nerve together with the ganglion from which it arises may be observed in a single cross section. Such a section, through the mandibular segment at Stage XI-XII, is represented in figure 50.A. Here the laterad extensions of the ganglia—composed of ganglion cells—are seen in process of being directly transformed into the lateral nerves innervating the mandibles. It will also be noted that the clefts in the lateral cords, later filled by nerve fibres, are in this section continuous with hollow spaces traversing the centre of the lateral nerves, which spaces are also, at later stages, filled by nerve fibres. Caudad of the gnathal region the lateral nerves are much longer and relatively more attenuated, and their development consequently more difficult to follow.

The median cord, up to Stage XI-XII (Figs. 49A and B), has altered little histologically from its condition at Stage VII, being for the most part still composed of rather long prismatic cells with large and clear nuclei. Its cells nevertheless are slowly increasing in number by mitotic division and decreasing in size. These divisions are equal, and appear rather inconstant in direction. In figure 50B a mitotic spindle is seen, directed obliquely. The size of the nuclei of the intraganglionic portions of the median cord, after their severance from the hypodermis, continues to decrease, while at the same time they lose their characteristic elongated form, until at Stage XIV their appearance is the same as that of the ganglion cells of the lateral cords, having undergone a similar differentiation into functional ganglion cells.

The intersegmental sections of the median cord, as already stated, remain united to the hypodermis, except in the gnathal segments and in the fifteenth, sixteenth, and seventeenth segments (Fig. 51B). In these the median cord is taken up entire, the intersegmental portion being also severed from the hypodermis and contributing to the formation of the ganglia of these segments, which are fused to form compound ganglia prior to hatching. In the gnathal region, however, a slight thickening of the ventral hypodermis marks the boundary between the rudiments of the mandibular and first maxillary ganglia. In the remainder of the trunk region the intersegmental sections of the median cord separate from the rest and form processes of the hypodermis, whose form, prior to hatching, is illustrated by figures 50C and E. At the time of hatching (Stage XV) the larva elongates, involving a

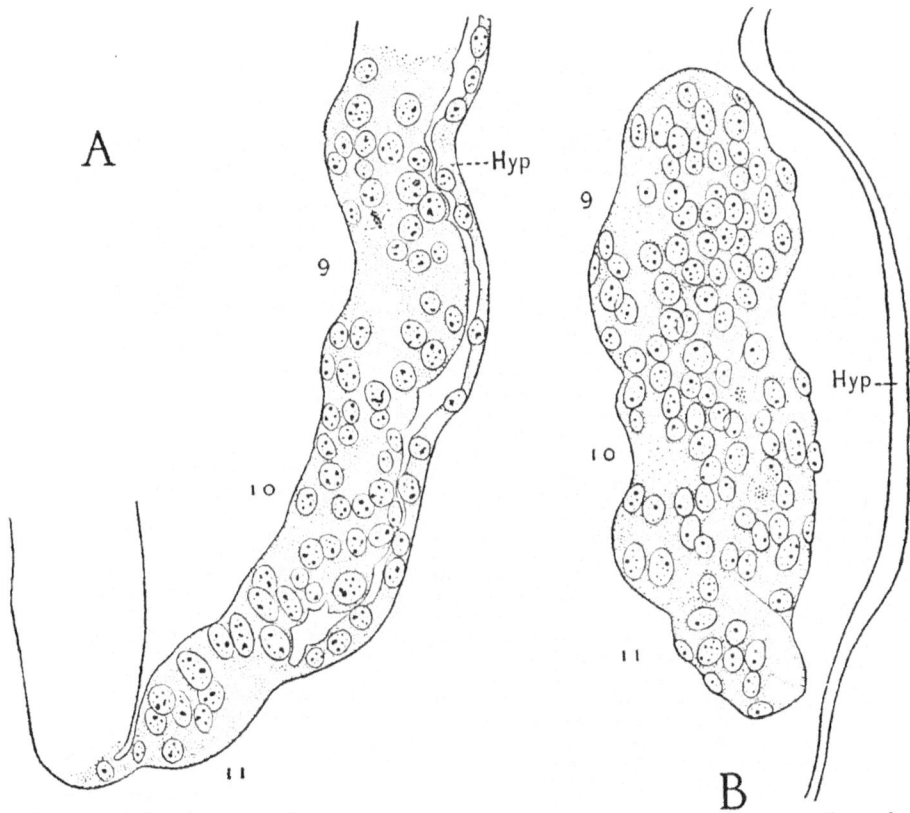

FIG. 51. Median sagittal sections through the last three ganglia of the ventral cord, corresponding to the 9th, 10th, and 11th (true) abdominal segment. A, Stage XI-XII, showing the ganglia separating from the hypodermis of the 9th, 10th, and 11th abdominal segments. B, Stage XV, showing two evident ganglia and the vestiges of a third, fused into a compound ganlion, x 600.

stretching in the longitudinal axis of all parts of the trunk. The effect of this process upon the intersegmental sections of the median cord is a decrease in their height. At Stage XV they have consequently the appearance of being mere transverse folds of the ventral hypodermis (Figs. 37B and D). The number of ganglia (or pairs of ganglia) in the ventral chain of the honey bee is seventeen. Of these the first three are fused to constitute the suboesophageal ganglion, and the last three similarly united to form a compound ganglion, which in the young larva (Fig. XV) is an oblong mass lying beneath the hind-intestine. This mass, in sagittal section (Fig. 51B) is seen to be subdivided by constrictions into three successive swellings, the anterior two of which are of nearly equal size while the third is much smaller.

The first two of the three swellings each possess two evident commissures, clearly demonstrating that these two swellings represent distinct ganglia, the third swelling on the other hand, shows no evidence of commissures. In favorable transverse sections, however, evidences of a few commissural fibres may be seen in this swelling also. The best evidence for its title to be considered as representing a pair of ganglia is obtained in sagittal sections of Stage XI-XII (Fig. 51A). Here the terminal swelling, like the two preceding it, is plainly seen to be derived from a distinct trunk segment, the seventeenth (the eleventh of the trunk), which is marked off by a well defined constriction. There are therefore three pairs of ganglia represented in the terminal swelling of the ventral cord, the last pair being rudimentary.

In the embryos of insects generally either sixteen or seventeen pairs of ganglia are recognizable; in the latter case the last pair is more or less rudimentary. Among the Hymenoptera Grassi (1884) in the honey bee found sixteen pairs; Graber (1890) in *Hylotoma* found sixteen pairs and the evident rudiments of a seventeenth pair; Carrière and Bürger (1897) in the embryo of *Chalicodoma* found the rudiments of seventeen pairs of ganglia. The investigators last named state that the rudiment of the last (seventeenth) pair of ganglia is somewhat shorter than the others. In figure XLV of their paper is a representation of the posterior end of an extremely young larva, showing clearly the last three pairs of ganglia of the ventral cord. These are here seen to be subequal in size, the seventeenth pair appearing to be nearly if not quite as large as the other two pairs.

The origin of the lateral cords in insects is fairly well established. In every insect embryo thus far studied they arise, as Hatschek (1877) discovered, from the inner layers of the primitive swellings. Wheeler (1891, 1893) called attention to the rôle played by the neuroblasts in the development of the lateral cords. Subsequent investigators of the development of the nervous system of insects have uniformly observed similar teloblastic cells whose behavior corresponds, with the exception of minor differences, to the account given by Wheeler.

At the end of the embryonic period the neuroblasts undergo

degeneration in the Orthoptera (Viallanes 1891, Wheeler 1893, Heymons 1895) and also in the Dermaptera (Heymons 1895). In *Donacia,* according to Lecaillon (1898), they simply disappear. Bürger (Carrière and Bürger 1897), says   (p. 370) in regard to *Chalicodoma:* "I do not think that the cells I have designated neuroblasts degenerate. Later they can not be distinguished from the cells which they have produced." In the honey bee the neuroblasts do not degenerate before hatching, since at Stage XV they are conspicuously visible at the periphery of the ganglia (and brain as well) and are frequently dividing mitotically (Figs. 37A, 37C, *Nbl,* see also Fig. 41).

The development of the median cord, unlike that of the lateral cords, appears to differ considerably in different insects. That it is derived from a median strip of the ventral ectoderm, forming the floor of the neural groove, seems to be at least certain. The ultimate fate of this strip is less uniform. All investigators of the subject—with the exception of Wheeler—agree with Hatschek (1877) that the intraganglionic sections of the median cord contribute at least a large part—if not all—of the median portions of the ganglia, including the transverse commissures. To this opinion, however, Wheeler (1893) takes exception. This investigator, while admitting that in *Xiphidium* the intersegmental regions of the median cord—the progeny of the median neuroblasts—are taken up into the central portions of the ganglia to form functional ganglion cells, does not believe that the median cord cells in the intrasegmental regions became ganglion cells, but believes that they are used up in the formation of neurilemma. In the same group to which *Xiphidium* belongs, the Orthoptera, Heymons (1895) later found that the anterior and central median gangliomeres were actually formed by the median cord, much as in other insects. With this exception, the differences in regard to the development of the median cord center principally about the fate of the intersegmental (interganglionic) sections. Hatschek (1877) stated that these remained in connection with the ectoderm and contributed nothing to the ganglia.

Graber (1890) found that in *Melolontha* the median cord was separated from the hypodermis throughout its entire length, but that the intersegmental portions later divide transversely, each half being then drawn cephalad and caudad respectively into the ganglia adjoining. In *Hydrophilus, Lina* and *Stenobothrus* on the other hand the median cord was not observed to separate intersegmentally from the hypodermis.

Wheeler's account of the development of the nerve cord of the grasshopper *Xiphidium* has already been outlined at the beginning of this section. The interganglionic sections of the median cord were each represented by a deep invagination, forming a part of the floor of the neural groove, and also by a median neuroblast, the median cord neuroblast. The former, the ectodermal invagination, produced no nerve tissue, but remained in connection with the hypodermis and later formed, in the thoracic region, the *furcae,* apodemes for the attachment of muscles; in the abdomen these invaginations also occurred but persisted only for a short time and then disappeared. The products of the median cord neuroblasts on the other hand become displaced cephalad and contributed to the formation of the posterior median neuromere.

Heymons (1895) in his researches on the Dermaptera and Orthoptera has virtually confirmed Wheeler's results, as concerns the interganglionic portions of the median cord. Heymons however found instead of one interganglionic neuroblast, several of these cells. In *Lepisma* Heymons (1897) found that a continuous median cord was set free from the hypodermis and present in the newly hatched nymph, extending the entire length of the ventral cord.

Carrière and Bürger's (1897) statements concerning the fate of the median cord in *Chalicodoma* are contained in the following paragraphs (pp. 371-372) : "My investigations essentially confirm the account given by Heymons (1895). Nevertheless I have not been able to determine, that the floor of the neural groove becomes split up into a dermatogenic and neurogenic layer. According to my observations all of its cell material goes to form the median cord, while its covering is produced by the union of the hypodermis formed in the region of the primitive swellings.

"The complete sundering of the median cord from the hypodermis takes place about the end of development. Its intraganglionic portions separate from the hypodermis at about the same time as the delamination of the lateral cords, its interganglionic portions remain in connection with it, after the nerve fibres have become evident in the lateral cords."

Escherisch (1902, 1902a) has given a very complete and interesting account of the development of the median cord in *Musca,* in which the conditions recall those found in *Lepisma.* In *Musca* a continuous median cord is separated from the ectoderm. Within the limits of the ganglia the median cord contributes their median portions, as in other insects; in the interganglionic regions it presents swellings of considerable size, one being situated directly caudad of each ganglion. From each of these swellings, near its posterior end, a pair of lateral processes are given off which extend to the neighborhood of the stigmata. These processes are regarded as nerves. Escherisch points out the close resemblance which the median cord bears to the unpaired median nerve described by several investigators of the anatomy of insects. This discovery, as well as that of Heymons (1897), shows that in some insects the median cord may form a more or less continuous median nerve. The presence of such a nerve in *Lepisma* suggests that its occurrence represents a primitive condition.

Hirschler (1909) in the chrysomelid beetle *Donacia,* has in part confirmed Graber's statements regarding *Melolontha,* since he finds that the median cord is at first completely severed, throughout its extent, from the hypodermis, but that afterward the intersegmental sections are added to the ganglia, forming in each the posterior median gangliomere. In their fate these interganglionic sections correspond to the median cord neuroblasts and their products in the Dermaptera and Orthoptera.

In the development of the lateral cords it is evident that the honey bee conforms to the general rule and that such differences as it presents are of relatively minor importance. In the origin and fate of the median cord it conforms to Grassi's account and is also similar to many of the Coleoptera, as, for example, *Hydrophilus* (Graber 1890), in so far as the median cord is not developed in the interganglionic regions, but remains here united with the hypodermis. These interganglionic spaces nevertheless are, up to the time of hatching, of very slight extent in an anteroposterior direction, as figure 50E shows; moreover the anterior and posterior commissures are close together, so that it is obvious that the anterior and posterior median gangliomeres, as well as the central gangliomere, are formed from the median cord.

B.   The Brain

The rudiments of the brain, the procephalic lobes (see p. 100), appear at Stages VI-VII. These together form a heart-shaped expansion of the anterior end of the germ band which at its widest part embraces a trifle over two-thirds of the diameter of the egg. The antero-lateral margin of each lobe is rounded, and the two lobes are separated from one another at the anterior end of the germ band, by a median notch or indentation. Caudad the lobes narrow rather gradually to join the remainder of the germ band. The anterior limit of the procephalic lobes is at this stage slightly ventrad of the cephalic pole of the egg.

The structure of the procephalic lobes at Stage VII is shown in figures 29, 30, and 31. As these show, the lobes (*ProL*) are composed of a single thick layer of long prismatic cells which rises slightly above the level of the surrounding blastoderm, (Fig. VII). On comparing the ectoderm of the procephalic lobes with that of the neurogenic area of the trunk it is at once evident that the ectoderm of the lobes is of much greater thickness, particularly in their anterior half (Figs. 30, 31). The nuclei of the cells composing the procephalic lobes are similar to those of the neurogenic area of the trunk ectoderm both in size and clear appearance.

In cross sections through the posterior region of the protocerebral lobes, caudad of the point where the anterior mesenteron rudiment comes to the external surface, there is seen on each side a group of cells (Fig. 31, *Hyp*) whose nuclei have the small size and dense appearance characteristic of the dermatogenic ectoderm bordering the neurogenic area of the trunk. By following this group caudad it is found to be actually continuous with this portion of the ectoderm, which may therefore be conceived as sending a tongue-like prolongation forward on each side into the neurogenic portion of the protocerebral lobes. The fate of this group of cells is uncertain, but its position suggests that it may represent the antennal rudiment, which at the following stage

(VIII) is actually composed of cells of just this character (Fig. 52, *Ant*). If these groups of small cells represent the antennal rudiments, then the antennae originate caudad of the future mouth, since its location on the germ band is marked by the point

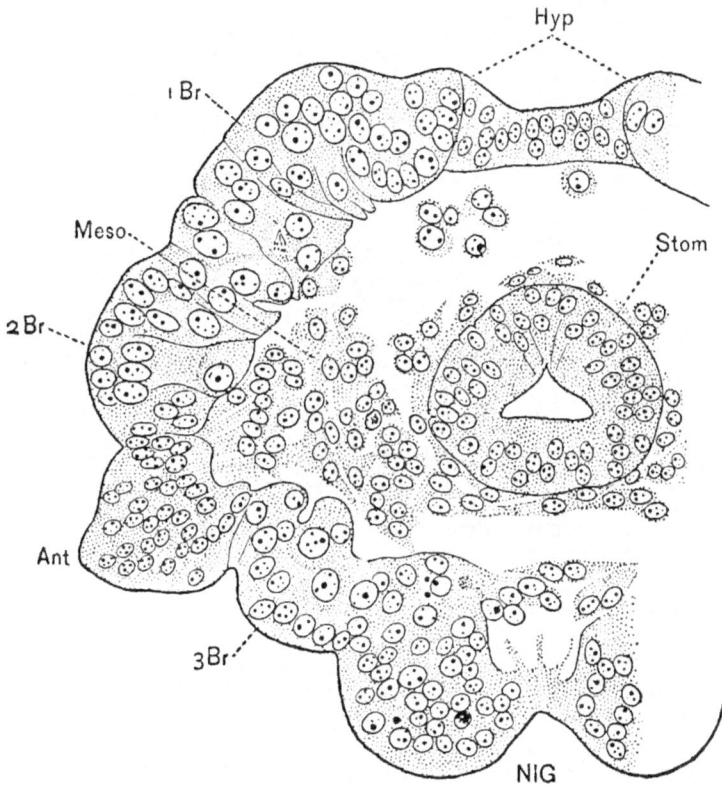

FIG. 52. Transverse section through the head of an embryo, Stage IX, showing the three principal divisions of the brain (*1Br, 2Br, 3Br*). and the antennal rudiment (*Ant*), x 387.

where the cells of the anterior mesenteron rudiment come to the surface.

At Stage VIII the procephalic lobes have undergone considerable change both in position and form. The entire germ band has increased in length, thereby bringing its anterior end around the cephalic pole of the egg, so that the procephalic lobes are now curved in a semicircle, their anterior ends lying on the dorsal side of the egg and directed toward the caudal pole. The procephalic lobes have meanwhile become subdivided into three lobes, the three pairs constituting respectively the proto-cerebrum, deutocerebrum and tritocerebrum, each of which rises above the external surface of the egg as a more or less evident swelling. Of these the protocerebral lobes (Figs. VIII, VIIIa, *1Br*) are the largest. They are relatively flat with a rounded external margin and have previously been described as resembling a saddle in shape. The deutocerebral lobes (*2Br*) are more convex, do not extend so far laterad, and are each tipped by a small papilliform projection, the antennal rudiment (*Ant*). The tritocerebral lobes (*3Br*) are small hemispherical elevations situated on a line with the posterior border of the stomodael depression and somewhat farther apart than are the neural ridges to which they are joined. They are in fact nothing more or less than the much discussed "second antennae" of Bütschli and the "transitory anterior appendage" of Grassi (see p. 109). A section transverse to the long axis of the egg and passing through these three pairs of lobes are shown in figure 52.

On each side are seen the papilliform antennae (*Ant*), immediately dorsad of which are the deutocerebral lobes (*2Br*). Above these and divided from them by a slight depression are the anterior ends of the protocerebral lobes (*1Br*) which extend almost to the mid-dorsal line, being here joined together by a bridge of small dermatoblastic cells (*Hyp*). Ventrad of the antennae are the tritocerebral lobes (*3Br*), which are clearly seen to be essentially similar to the remainder of the ganglionic swellings of the neural ridges. Between these lobes, on the ventral side, is a pair of high rounded elevations separated by a rather shallow median depression. These elevations are merely the extreme anterior ends of the neural ridges, which, at this and the stage next succeeding terminate anteriorly in a pair of rounded swellings situated just behind the mouth, and which are very evident in figures IX and X.

During Stage X the external elevations marking the two anterior divisions of the brain become less evident, the outer surface of the head becoming relatively smooth in contour. The swellings forming the tritocerebrum tend to lose their prominence and at Stage X (Fig. X) have almost disappeared. This flattening out of the external evidences of the rudiments of the neuromeres of the brain takes place at the same stage and corresponds with the flattening out of the neural ridges in the trunk, the rudiments of all of the neuromeres now rising above the internal instead of the external surface of the ectoderm.

The lateral aspect of the head at Stage X shows little evidence of the two anterior brain neuromeres. Its surface appears hemispherical, smooth and unbroken except for the button-like antennae. The dorsal aspect of the head however enables one to gain a conception of the general form of the protocerebral lobes at this stage. This is illustrated by Fig. 53, taken from a camera draw-

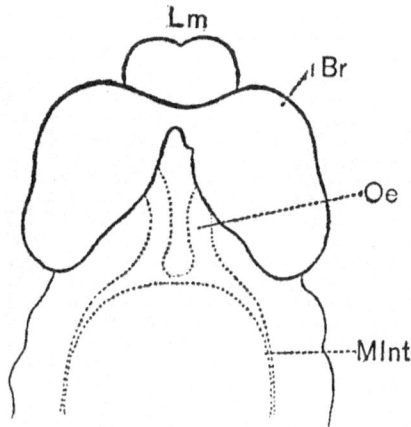

Fig. 53. Dorsal view of embryo, Stage XII, showing outlines of protocerebrum (*1Br*), x 112.

ing. The protocerebral lobes are now seen to be oval in outline, their smaller ends directed backward and outward, their long axis diverging at any angle of about fifty degrees.

The prismatic cells composing the ectodermal thickening which constitutes the future brain, at Stage VIII are precisely similar to those of the corresponding neurogenic area of the trunk. During Stage IX and X the brain rudiment undergoes a differentiation corresponding with that of the ventral cord. Near the inner surface of the ectoderm of all three segments of the brain appear the clear and rounded nuclei of neuroblasts (Fig. 54, A and B, *Nbl*). The future history of these cells appears to be essentially similar to those of the ventral cord, although their study in the brain is much more difficult than in the ventral cord because of the way in which the brain is bent about the anterior end of the egg. Nevertheless, in favorable sections the same phenomena may be observed. Figure 54A shows the unequal teloblastic division of a neuroblast in the rudiment of the deutocerebral segment, and also two ad-

jacent neuroblasts, one of which is seen accompanied by two of
its progeny. Figure 54B taken from the tritocerebrum shows three
parallel rows formed by the neuroblasts and their descendants.

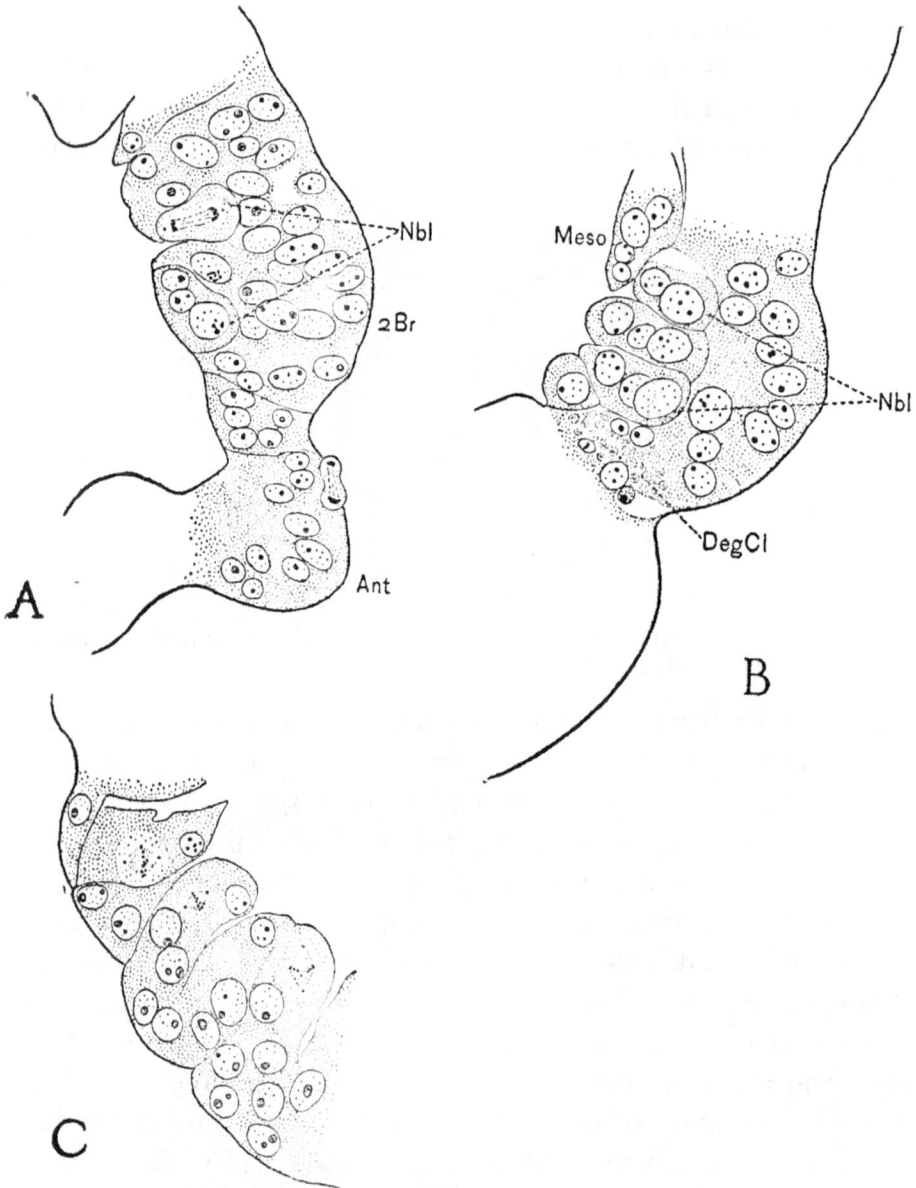

Fig. 54. Parts of sections through the brain of embryos, to show the
neuroblasts (*Nbl*) and their products. A is from a transverse section of
Stage IX, passing through the antennal rudiment (*Ant*), and shows
the rudiment of the deutocerebrum (*2Br*). B is from a sagittal section
of Stage IX, passing through the tritocerebrum. C. includes a portion
of the dorsolateral surface of the head of an embryo of Stage X-XI, and
shows small spindles of two different sizes, x 600.

Figure 54C taken from the protocerebrum shows three spindles, the two larger being plainly those of neuroblasts, while the smaller is that of a daughter cell of the first generation.

A minor point in which the histogenesis of the brain differs from that of the ventral cord is the greater irregularity of the groups formed by the neuroblasts and their progeny. In the ventral cord these form more or less regular rows, lying for the most part in the transverse plane of the embryo, while in the brain the different neuroblasts and their progeny often appear to form a confused mass. This is especially true of the earlier stages, and is at least in part to be ascribed to the varying planes, with regard to the morphological long axis of the embryo, in which the brain is intersected. In the later stages more or less regular pyramidal groups with a neuroblast at their outer and larger ends are frequently seen. Some of these are shown in figure 56.

Like the neuroblasts of the ventral cord, those of the brain persist until after the hatching of the young larva. Probably they remain active much longer than this. Sections through the brain of a larva about two days old show cells with large clear nuclei situated about the periphery of the brain and having all the appearance of the embryonic neuroblasts, except that they are larger. Two of these larval cells were observed undergoing an unequal and tangential division precisely like that of the embryonic neuroblasts.

At Stage X, when the brain has nearly reached its ultimate embryonic dimensions, it begins to separate from the hypodermis. This takes place in the same way as in the ventral cord, and the hypodermis has for the most part the same character, being thin and made up of flat cells. The deutocerebrum and tritocerebrum separate from the hypodermis first. The protocerebrum separates from the hypodermis more slowly, the separation beginning at the mesial border of these lobes at Stage XI, and slowly progressing laterad, being completed as Stage XII (Figs. 52 and 55).

The nerve fibres make their appearance at Stage XI (Fig. 55), appearing simultaneously in the three divisions of the brain. Their mode of development is precisely the same as in the ganglia of the ventral chain, the formation of the nerve fibres being preceded by the appearance of clefts beneath the innermost row of cells of

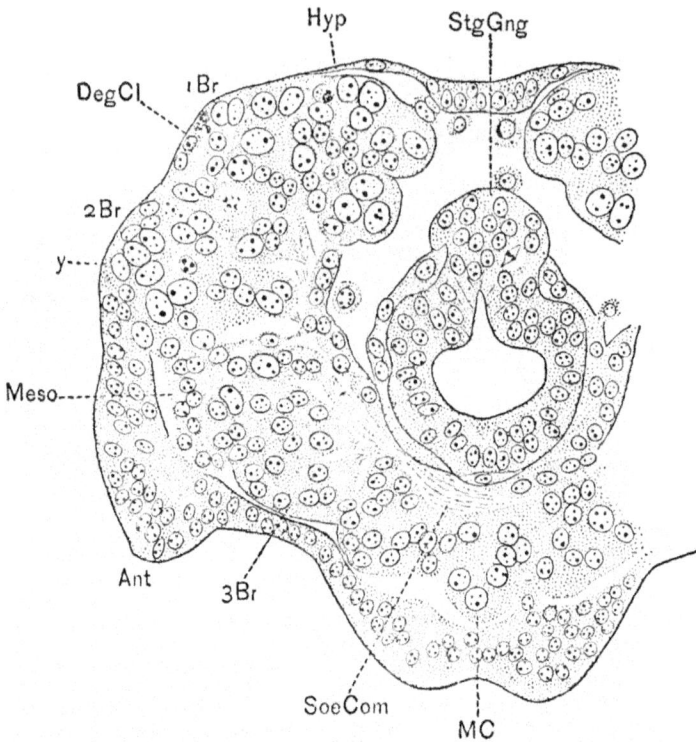

FIG. 55. Transverse section through head of embryo, Stage XI-XII, showing the formation of the suboesophageal commissure (*SoeCom*) and the stomatogastric ganglion (*StgGng*), x 387.

the brain rudiment (Fig. 55). These clefts then widen out and become occupied by the fibres (Fig. 56).

While the greater portion of the brain is formed by neuroblasts, parts of both the deuto- and protocerebrum are not formed in this way. These exceptions include the optic lobes, whose development will be described later, and a pair of small groups of prismatic cells, situated, when first seen, one on each side, just above the base of the antennal rudiment. These groups of cells become plainly noticeable at Stage XI as a subspherical cluster of large prismatic cells (Figs. 55, and 57, *y*). They apparently do not divide, at least up to the time of the hatching of the larva, but are covered over by the hypodermis derived from the base of the antennal rudiment as shown in figure 57. At Stage XV they are readily recognizable as a spherical group of relatively large cells lying embedded in the deutocerebrum close to its outer surface (Fig. 41, *y*).

The optic lobes of the insect brain constitute the first, and with

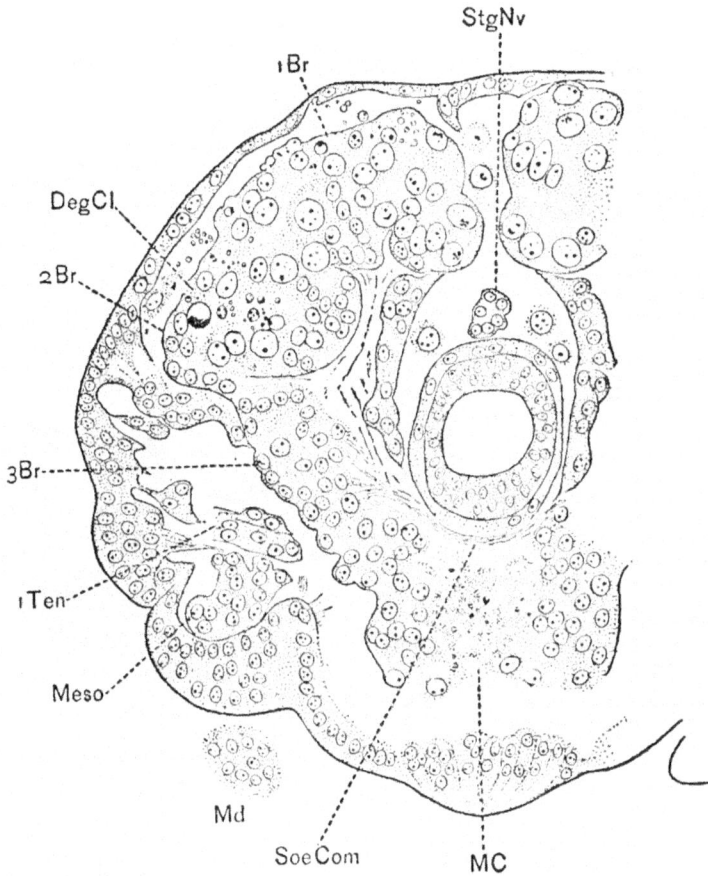

FIG. 56. Transverse section through head of an embryo, Stage XIII-
XIV. This section passes just caudad of the antennal rudiments, x 387.

reference to the rest of the brain, the outermost of the three
divisions of the protocerebral lobes (see pp. 116-117). Since in the
honey bee these lobes are, with reference to the poles of the egg,
directed caudad and laterad, the optic lobes may be similarly
described as derived from the lower and posterior border of the
protocerebral lobes, although this is actually their upper (dorsal)
and anterior border. As in other insects then development differs
from that of the remainder of the brain.

At Stages VIII and IX the ectoderm destined to form the optic
lobes is not distinguishable from the other neurogenic ectoderm
of the brain rudiment. During the succeeding stages the greater
part of the latter becomes transformed into neuroblasts and ganglion cells and the regular palisade-like arrangement of its cells
altogether disappears. This does not occur in the region of the

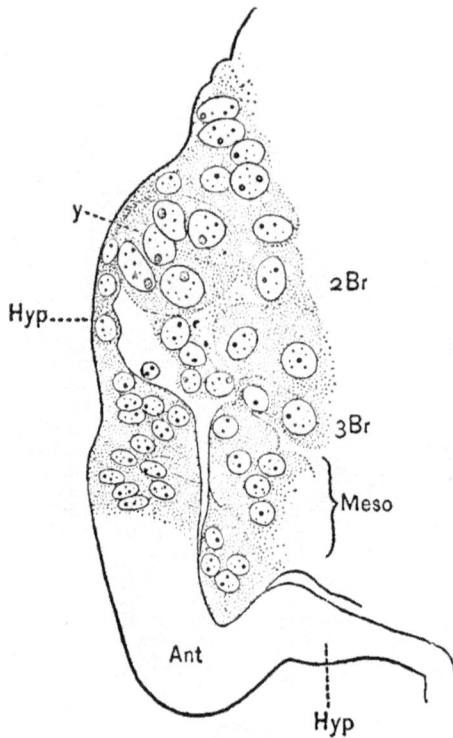

FIG. 57. Part of a transverse section of Stage XI-XII, passing through the antennal rudiment (*Ant*), showing one of the pair of spherical cell groups of unknown significance (*Y*), which enter the deutocerebrum (*2Br*), x 600.

future optic lobes. Here the regular palisade-like arrangement characteristic of the neurogenic epithelium at earlier stages is preserved, the rudiment of the optic lobes appearing as a thick epithelial plate composed of slender prismatic cells with large nuclei (Fig. 58A, *OpL*). Only a limited number of neuroblasts are produced from this area. These, together with the progeny of adjacent neuroblasts form a mass of cells underlying all but the extreme lateral borders of the epithelial plate. On its external rounded margin the epithelial plate is sharply demarcated from the dermatogenic ectoderm immediately adjacent, which equals the epithelial plate in thickness and is composed of narrow prismatic cells of the smaller size characteristic of the dermatogenic ectoderm (Fig. 58A, *OpPl*). This dermatogenic layer becomes thin at its edges where it meets the optic lobes, the thin edge overlapping and covering their margins.

The epithelial plate destined to form the optic lobes is at Stages

XI and XII plainly seen to be marked off into two nearly equal parts by a narrow furrow which runs parallel to the outer margin of the protocerebral lobes. A similar furrow or depression on the opposite side of the epithelial plate combines with the first to form a constriction which separates the plate into two nearly equal parts, one of which is central, lying next to the second division of the protocerebrum, while the other is marginal or peripheral (Fig. 58A). This half projects laterad nearly free from the mass of cells constituting the adjacent parts of the protocerebrum.

During Stages XII-XIII the superficial furrow deepens into a cleft, the two halves of the epithelial plate simultaneously bending inward in such a way as to form an invagination of which the cleft is the lumen (Fig. 58B). The cells at the bottom of this invagination are considerably deformed, the lateral two-thirds of the epithelial plate being, so to speak, doubled up into a form resembling that of the letter U (Fig. 58C). The peripheral half of the epithelial plate together with the adjacent dematogenic ectoderm, forms a double fold constituting the lateral boundary of the invagination. During Stage XIII, the two layers of this double fold separate from one another, the outer dermatogenic layer gliding over the optic lobe to join the thin hypodermis formed from the adjacent parts of the brain by delamination (Fig. 58C). This thicker hypodermis overlying the optic lobes now constitutes the *optic plate,* from which the receptive portion of the compound eye is derived (Fig. 58A, B, and C, *OpPl*). The optic lobe of each side now appears in either transverse or coronal sections as an oval mass composed of a double layer of large columnar cells, and retains this appearance at least well into the larval period.

A conception of the superficial extent and direction of the invagination concerned in the formation of the optic lobes may be obtained from certain favorable tangential sections of the head. Such a section is represented in figure 58D. The double fold of neurogenic ectoderm forming the optic lobe (*OpL*) is at Stage XV seem to extend across the head at approximately right angles to the long axis of the protocerebral lobes. These two folds are continuous with one another at both ends of the slit-like lumen of the invagination, so that each optic lobe may be compared to

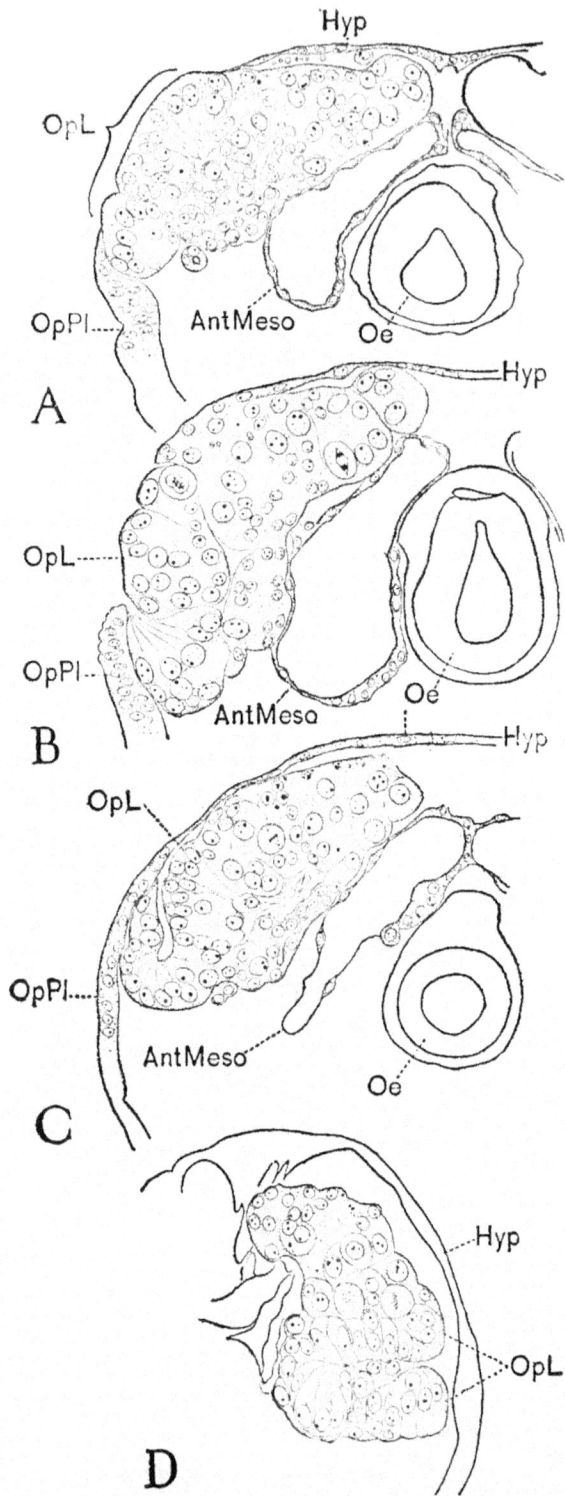

FIG. 58. Sections through the otic lobe of embryos. A and B, transverse sections intersecting the left optic lobe of two embryos of Stage XII; C, of Stage XIII-XIV. D, tangential section through one of the optic lobes of a newly hatched embryo (Stage XV), x 290.

an ordinary cup-like invagination greatly drawn out and flattened. At stages prior to hatching the long axis of the invagination is oblique to the long axis of the embryo, so that both coronal and transverse (Fig. 58A-C) sections intersect the long axis of the invagination. At hatching the dorsal flexure of the embryo is exchanged for a ventral flexure, and the head becomes turned ventrad, so that the long axis of the invagination is brought to a position approximately at right angles to the long axis of the young larva (Fig. 39).

The accounts of the formation of the optic ganglion and optic plates of insects differ to a considerable extent. In *Mantis* (Viallanes 1891) the optic ganglion is produced by "Cellules gangliogènes" or neuroblasts like the remainder of the brain, the overlying hypodermis splitting off from the optic ganglion to constitute the optic plate. In *Xiphidium* (Wheeler 1893) the optic ganglion and optic plate are formed by a simple separation of two cell layers, as is *Mantis,* but the inner layer is single and contains no neuroblasts. In *Forficula* (Heymons 1895) the process is much the same, but Heymons reports having observed a few neuroblasts in the layer forming the optic ganglion. In none of these forms is there any invagination of the ectoderm forming the optic plate. There is present however, an ectodermal invagination or proliferation just posterior to the embryonic optic lobe (Fig. 38, *Igl*), but this is outside of the limits of the latter and situated between it and the second protocerebral lobe, and is moreover purely hypodermal in character, so that—as will appear later—there seems to be no good ground for comparing this invagination, the intergangionic thickening of Wheeler (1893), with the invaginations of the optic lobe described by Patten (1889), as Viallanes (1891) and Wheeler (1893) have done. In the remaining accounts, which relate exclusively to the orders Coleoptera and Hymenoptera a conspicuous ectodermal invagination is concerned in the development of the optic lobe. Patten (1887) was the first to observe this in the case of the wasp *Vespa.* Here the rudiments of the eye are formed somewhat as in the honey bee, the optic plate lying at first laterad of the rudiment of the optic plate, and later extending mesiad over it. The separation of the optic plate from the optic lobe however takes place early, and the ectoderm forming the optic lobe is not folded to any extent, being merely bent inward to form a cup-shaped cavity. At the beginning of the process, the optic plate is connected with the optic lobe by a strand of cells which is said later to constitute the optic nerve. Nothing of this kind was seen in the honey bee.

In Patten's next paper (1889), on the eyes of the beetle *Acilius* a long curved slit-like invagination, or rather three such invaginations, almost continuous, were described as concerned in the formation of the optic lobe. One of these invaginations was considered as rudimentary. The other two seem together to correspond very closely to the invaginations forming the optic lobes of the bee in their form, situation and relation to the surrounding parts. As in the honey bee, the ectoderm destined to form the optic plate lies at first external to the future optic lobe.

Heider (1889) described a similar condition in *Hydrophilus*. Heider's text figure 5 representing a transverse section shows an infolding strikingly similar to 58B. All the relations are apparently the same as in the honey bee. Moreover the ectoderm from which the optic lobe is formed appears to also consist of a single layer of columnar cells.

All the cases thus far studied fall broadly into two classes: those in which the optic lobe is formed by delamination, and those in which it is formed by infolding. To the first class belong representatives of the order Orthoptera and the nearly related order Dermaptera. To the second class belong representatives of the Coleoptera and Hymenoptera. In this class the optic plate is formed from ectoderm lying outside of and immediately surrounding that destined to form the optic lobes, in other words the optic lobe and optic plate are formed from separate areas of the ectoderm, while in the first class they are formed from the same area. In the first class (Orthoptera, Dermaptera) the optic lobes undergo rather a complex series of changes in order to arrive at their ultimate or imaginal form. The postembryonic changes undergone by the optic lobes in insects belonging to the second class are unknown so that a basis for a definite comparison is lacking, and only a guess is possible as to the meaning of the different parts. A comparison of the optic lobes of the bee at Stage XV with those of *Mantis* or *Xiphidium* at a stage shortly prior to hatching suggests that the folding of the optic lobe in the bee—and in the Coleoptera—may correspond to the subdivision of the optic lobe in the forms first mentioned into the ganglionic layer and the external medullary mass.   This is however only a surmise.

Both the supra- and suboesophageal commissures have been compared by writers on the insect brain to the commissures of the ventral cord, and are supposed to be formed in the same way, that is, by a median ingrowth of the ectoderm comparable to the median cord.   Heymons has observed this in both commissures in *Forficula*.   In the honey bee the cellular portion of the supraoesophageal commissure is formed from a median thickening of the ectoderm of the head.   Figure 59B, taken from a transverse

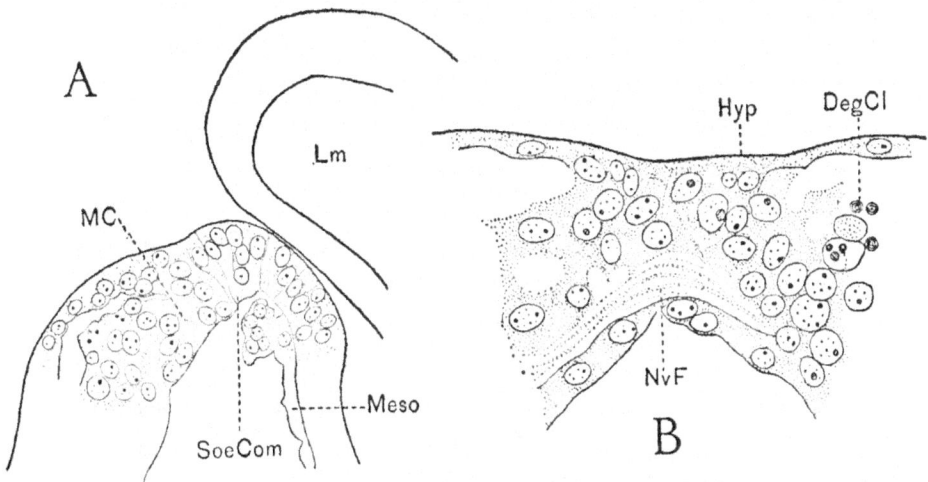

FIG. 59.   A, median sagittal section through the mouth of an embryo, Stage X, showing the suboesophageal commissure (*SoeCom*).   B, part of a transverse section of an embryo, Stage XIII-XIV, showing the formation of the supraoesophageal commissure, x 600.

section of Stage XIII-XIV sufficiently illustrates this point.   As in the commissures of the ventral cord it is the cellular portion principally which is furnished by the median ingrowth, the bridge of nerve fibres arising principally from the ganglion cells of the paired ganglia lying on each side of the mid-line.

The origin of the suboesophageal commissure, or even a part of it, from the ectoderm of the mid-line, is much less clear.   As figure 41, Stage XV, shows, the commissure (*SoeCom*) consists principally of an extremely thin band of nerve fibres connecting

together the two halves of the tritocerebrum, and is accompanied
by a few small cells, apparently ganglion cells. This band of
fibres is formed at Stage X-XI by processes of the ganglion cells
of the tritocerebral lobes (Fig. 55, *SoeCom*). These fibres lie in
a minute channel at the bottom of the reentrant angle formed by
the junction of the hinder wall of the oesophagus and the ventral
ectoderm and lie immediately above the extreme anterior end of
the median cord (Fig. 59A, *SoeCom*). During Stage XI after
the suboesophageal ganglion separates from the ventral hypodermis the suboesophageal commissure also leaves the ventral body
wall, although remaining in contact with that of the oesophagus.
During the stages succeeding, the hypodermis directly caudad of
the mouth grows out to form the lower lip or hypopharynx, increasing the space below the oesophagus and consequently the
distance of the circumoesophageal commissure from the ventral
surface. Up to Stage X the nerve fibres of the suboesophageal
commissure and the anterior end of the median cord, now an
integral part of the suboesophageal ganglion, remain in close apposition (Fig. 59A). At Stage XIII-XIV (Fig 45) these become
separated by a slight cleft traversed by a few muscle fibres (dilators of the pharynx) which pass from the median part of the
tentorium to the posterior or lower wall of the pharynx.

   In view of the close apposition of the fibres of the circumoesophageal commissure and the anterior end of the median cord, and
of the serial homology of the commissures of the ventral cord
with those of the brain, it is not impossible that the cells accompanying the fibres of the suboesophageal commissure are derived
from the ventral cord. This point could not however be definitely
determined, in spite of repeated efforts. While it may be true
that these cells are actually derived from the median cord, they
have rather the appearance of having migrated mesiad from the
two lobes of the tritocerebrum.

Summary

   In summing up the foregoing observations on the development
of the brain of the honey bee, with reference to their bearing on
the development of the brain of other insects, it may be said,
first, that in a broad sense they confirm the results of Viallanes,

Wheeler, Heymons and others, especially with regard to the division of the insect brain into three segments or neuromeres. Second, with regard to details, the brain of the bee shows several points of difference when compared with the few accounts available. These accounts relate principally to the development of the brain in Orthoptera, and the nearly allied Dermaptera, as exemplified by *Forficula*. In fact since the publication, in 1895, of Heymons' monograph, there has appeared no detailed account of the embryonic development of the insect brain. Even Carrière and Bürger (1897), in their otherwise quite complete account of the development of the mason bee, give a disappointingly brief account of the brain, concluding with the statement that "An extended account of the development of the brain may be omitted, since it would only serve to confirm the more recent investigations on this subject."

The points in regard to which the development of the brain of the bee differs from the Orthoptera and Dermaptera may be summed up as follows:

(1) The brain is flexed around the cephalic pole of the egg in such a way that the morphological anterior ends of the protocerebral lobes are directed toward the caudad pole of the egg.

(2) The cells of the second generation from the neuroblasts, instead of the first, form the definitive nerve cells. This is the same as in the ventral cord.

(3) The three subdivisions of the protocerebrum are not at first plainly marked off from one another and are never separated by hypodermal ingrowths.

(4) The optic lobes are formed, apparently independent of the agency of neuroblasts, by a deep invagination of the neurogenic ectoderm, which has no counterpart in the Dermaptera or Orthoptera, but seems to correspond more or less closely with an invagination of the optic lobe described by Patten (1889) in *Acilius* and Heider (1889) in Hydrophilus. Patten (1887) has also described an invagination concerned in the formation of the optic lobe in *Vespa* which seems to be similar to that in the honey bee.

C. The Stomatogastric System

The stomatogastric system, as in other insects is formed from the dorsal wall of the stomadaeal invagination. The rudiments of this system first make their appearance at Stage IX, but are more sharply marked at Stage X. At this stage three evaginations of the dorsal wall of the stomodeaum wall are visible (Fig. 60).

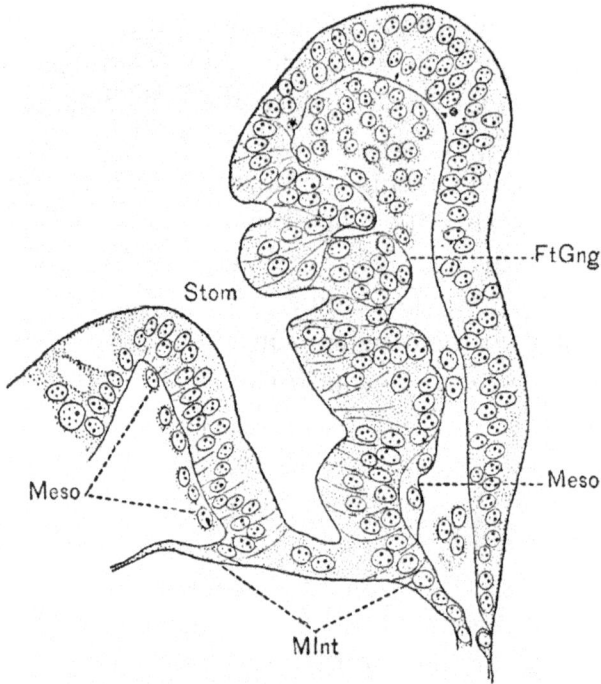

FIG. 60. Median sagittal section through the stomodaeum of an embryo, Stage X, showing the three evaginations of the dorsal wall of the stomodaeum which form the stomatogastric system, x 387.

On the lower surface of this wall these appear as sharply pointed clefts, on the upper surface as rounded swellings (Fig. 55, *FtGng*). The first (anteriormost) of these is relatively insignificant, while the other two are very noticeable. The first is situated not far from the end of the labrum, the second close behind the first, while the third is halfway between the second and the point where the oesophagus joins the mid-intestine. The first evagination liberates a few scattered ganglion cells which later are found accompanying the labral nerve, and then quickly disappears. The second forms a large ovoid mass of nerve cells lying just above the oesophagus and between the two halves of the brain, and is easily recognized as the *frontal ganglion* (Fig. 55, *FtGng*). This mass is at first hollow, with the cavity of the evagination extending up through it, but soon becomes solid. The third mass of

nerve tissue thus formed is lower and longer in an antero-posterior direction than the rudiment of the frontal ganglion, and is in
contact with the latter. This third mass forms the pharyngeal
ganglia of the stomatogastric system and possibly a part of the
frontal nerve connecting these ganglia with the frontal ganglion.

The frontal ganglion is known to arise from a median evagination of the dorsal wall of the stomodaeum in Orthoptera (Viallanes 1891, Wheeler 1893, Heymons), Forficula (Heymons 1895),
Coleoptera (Heider 1899), Hymenoptera (Carrière 1890) so that
it is safe to assume that this mode of origin is typical for the
insects. The accounts of Heymons and of Carrière and Bürger
of the origin of the frontal ganglia and its associated structures
are the most complete and circumstantial. In Forficula and several of the Orthoptera the stomatogastric system consists of the
frontal ganglion, close behind which is the elongate occipital
ganglion connected with the frontal ganglion by the short nervus
recurrens (stomatogastric nerve). From the occipital ganglia two
nerves run dorsad and caudad for a short distance to the paired
ganglia pharangea (pharyngeal ganglia), and these in turn send
out nerves which run to the posterior termination of the
oesophagus.

The stomatogastric system in Chalicodoma (Carrière and Bürger (1897) is essentially similar to that of the honey bee, consisting simply of a frontal ganglion, a recurrent (stomatogastric
nerve) and a pair of pharyngeal ganglia.

On comparing the stomatogastric system of these two representatives of the Hymenoptera with that of Forficula and Gryllus
it becomes evident that the principal difference between them
lies in the apparent absence of the occipital ganglion in the representatives of the Hymenoptera. In both the mason bee and the
honey bee the stomatogastric nerve is surrounded by ganglion
cells for some distance caudad of the frontal ganglia, and this
part of the nerve may therefore readily be homologized with the
occipital ganglion, as Carrière and Bürger have already stated.

The mode of origin of the stomatogastric system in Forficula
and Gryllus differs slightly from that in the honey bee. Three
evaginations are found on the dorsal wall of the stomodaeum.
From the first of these is formed the frontal and occipital ganglia,

from the second the pharyngeal ganglion, and from the third the nerves passing caudad from the pharyngeal ganglia. The first evagination in *Forficula* therefore corresponds apparently to the second in the honey bee, the second to the third, while the last is wanting.

D. Neurilemma

The neurilemma, which at Stage XV constitutes a thin cellular membrane covering the external surface of the nervous system (Fig. 37A, B and C, *Nlm*), first becomes evident in the ventral cord at Stage XI-XII. At this stage, it will be remembered that a considerable shifting and rearrangement of the cells of the lateral cord takes place as shown in figure 50A and B. A large number of the daughter cells of the neuroblasts have, in the intra-ganglionic regions, shifted laterad, while others appear on the ventral surface of the lateral cords where they form a single layer. A similar layer is already present at the dorsal surface. From these superficial cells the neurilemma is evidently derived. In figures 50A and B, on the ventral side of the lateral cords, the neurilemma may be seen to be in process of formation, certain of the superficial cells having already assumed a flattened form (*Nlm*). The development of the neurilemma on the dorsal side of the ventral cord is less readily observed, owing to the number of small cells crowded together in this region, but it seems reasonable to assume that its origin is the same here as on the ventral surface. The cells of the neurilemma at later stages (XI-XV) vary much in size, indicating division subsequent to their assumption of a superficial position.

In regard to the origin of the neurilemma in insects there has been a considerable difference in opinion. Nusbaum (1883, 1886) and Korotneff (1885) traced its origin to wandering blood cells. Wheeler (1893) believed that it arose from the intraganglionic sections of the median cord. Heymons (1895, p. 45) states that in *Forficula* "The outer neurilemma apparently arises from cells which during the segregation of the neuroblasts from the dermatogenous layer were separated off from the latter."

In the honey bee it seems highly improbable that the neurilemma arises from blood cells, since the cells which are destined to form the neurilemma are when first evident—at Stages XI-XII—

so closely associated with the ganglion cells of the ventral cord and so like them in appearance. It is still less probable that they owe their origin to the median cord since at the time they are first evident the median cord is just separating from the hypodermis, and moreover its cells are still relatively large and preserve their original elongate form. There accordingly remain two possibilities: either the neurilemma cells are, as Heymons suggests, split off from the dermatogenous layer; or else they are merely transformed ganglion cells. The former view has much to commend it, from a theoretical standpoint and, as far as the writer's observations go, can not be totally excluded. The evidence at hand however seems to favor the latter view. Prior to Stages XI-XII the line of separation between the lateral cords and the future hypodermis is coincident with the outer boundary of the neuroblasts, in other words there is no direct evidence that any considerable number of dermatogenous cells are split off from the peripheral layer; on the other hand at the time the lateral cords are definitely split off a rearrangement of the cells takes place in the intraganglionic regions of the lateral cords, at which time a number of small cells appear on their ventral surfaces. These cells closely resemble the ganglion cells. It seems probable therefore that in the honey bee the neurilemma owes its origin to the ganglion cells themselves.

E. Corpora Allata

The origin and development of the corpora or ganglia allata in the honey bee corresponds very closely to the account given by Heymons (1895) for *Forficula*. In both insects these bodies arise as ectodermal ingrowths located in front of the bases of the first maxillae. In the honey bee this is also the location of the tubular invaginations which form the apodemes, of the adductor muscles of the mandibles. Each of these ingrowths at Stage X has already become a long hollow finger-like structure curving dorsad and caudad. The mouth of this invagination is wide, invading the base of the maxilla on its anterior and lateral sides. At Stage X there may be found near the outer and caudal end of each of these openings a small tubular invagination extending mesiad into the base of the maxilla. These invaginations are the rudi-

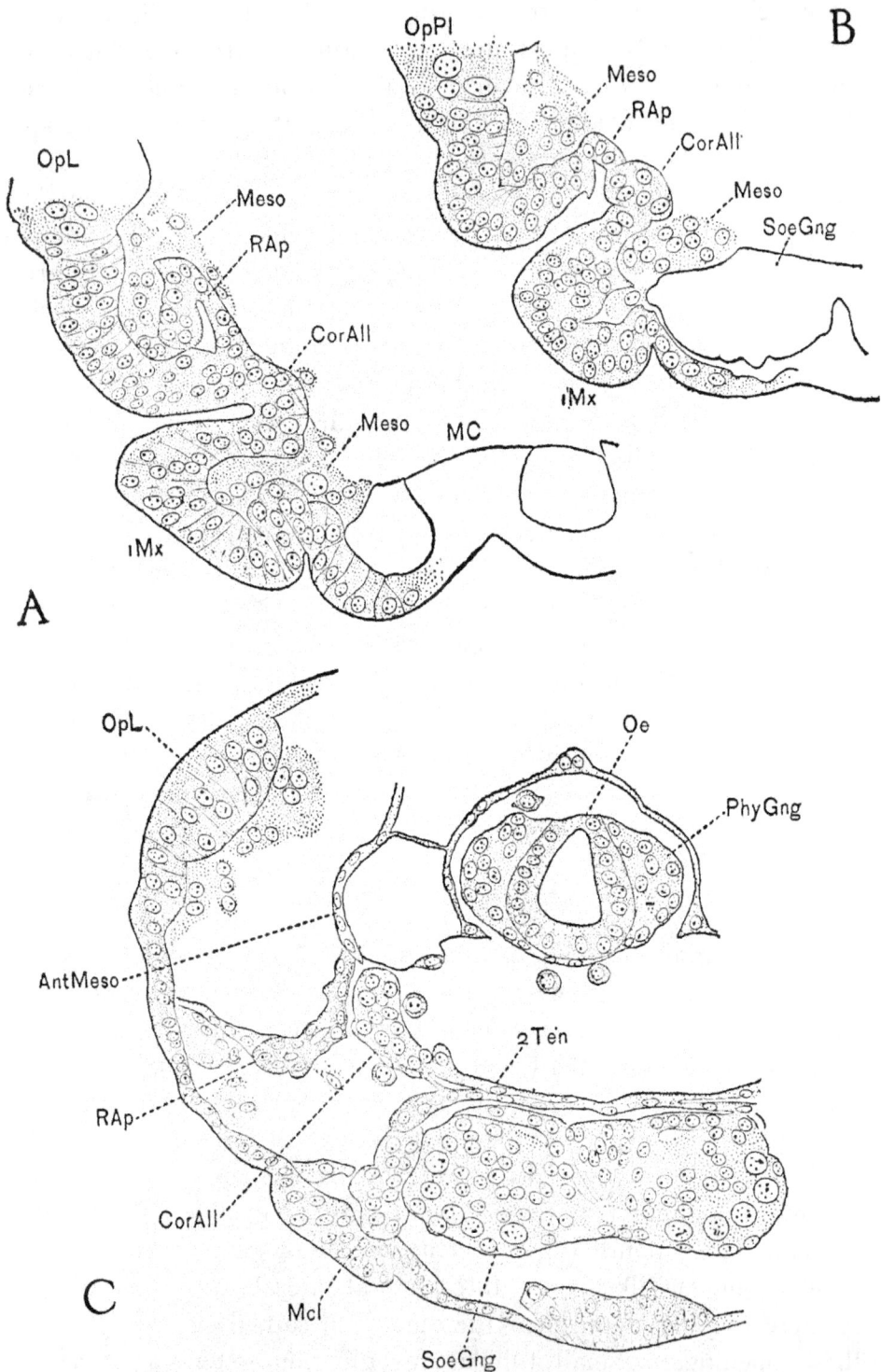

Fig. 61. Transverse sections through the head of three embryos of different stages, to illustrate the development of the corpora allata, (CorAll). A, Stage X; B, Stage XI; C, Stage XIII-XIV, x 387.

ments of the corpora allata (Fig. 61A, *Cor.All*).   At the next
stage (XI) the lumen of these invaginations has nearly disappeared and the corpora allata now appear as irregular cellular
outgrowths springing from the mesial side of the mandibular
apodemes (Fig. 61B). These cell masses are next constricted
off from the apodemes, but still remain in close contact with them.
Meanwhile the corpora allata are carried dorsad by the growth
of the mandibular apodemes and at Stage XII lie between the
latter and the inner ends of the posterior arms of the tentorium
where these join the anterior arms to form the central body. At
Stage XIII the corpora allata become attached to the ventro-
lateral angles of the antennal somites (Fig. 61C) and soon after
lose their attachment to the mandibular apodemes. This is their
position at the time of hatching and for at least a considerable
period of the life of the larva. The corpora allata acquire their
characteristic globular and compact form during the final stages
of embryonic development.

Heymons (1895) was the first to call attention to the bodies
named by him "ganglia allata," describing their development in
some detail in *Forficula,* and more briefly in *Gryllus.* Three years
later Carrière and Bürger (1897) found that in the mason bee
(*Chalicodoma*) the ganglia allata have the same origin as in
Forficula, but these bodies do not, however, later fuse to form a
median body as they do in *Forficula* but remain attached to the
ventro-lateral margins of the antennal coelomic sacs, as in the
honey bee. These investigators expressed a doubt as to the nervous nature of the "ganglia allata," since no nerve fibres could
be observed within them. In 1899 Heymons published a brief
paper on the structure and development of the corpora allata
of a walking stick (*Bacillus*). In this form these bodies lie one
behind the other, above the oesophagus and caudad of the pharyngeal ganglia, from which they receive nerves. In this insect the
corpora allata have the form of capsules within each of which
is a concentric layer of chitin secreted by the capsular wall. Their
development was found to be the same as in *Forficula.*

The function and homology of these bodies is unknown. Heymons (1899) surmised that they had a static function, but exci-
sion of the corpora failed to produce any disturbance in the locomotion of the insect (*Bacillus*). Janet (1899a, 1900) discusses at

some length the probable homology of the "corpora incerta" and arrives at the conclusion that these bodies probably represent the tracheal rudiments of one of the cephalic segments and that they furnish material for the tracheae of the head. No evidence to support this view has been seen in the honey bee. On the contrary the corpora allata and the tracheae develop independently of one another. In the honey bee the tracheae supplying the head are derived from tracheal invaginations located on the second maxillary segment (see p. 172).

A possible clue to the homology of the corpora allata appears in an interesting observation, apparently hitherto unnoticed, by Toyama (1900) on the embryo of the silkworm (*Bombyx*). In the section devoted to the endoskeleton of the head Toyama states: "In the mandibular segment three pairs of invaginations take place; the most anterior (between the labrum and the mandible) becomes the first tentorium, the second pair gives rise to the seat of the extensor mandibulae, while the last becomes the flexor mandibulae and salivary gland."[17] The flexor mandibulae and salivary gland therefore arise from a common invagination, which Toyama's figure (woodcut Fig. 1) shows to be situated at the posterior margin of the base of the mandible, corresponding quite closely to the point of origin of the flexor mandibulae and corpora allata in the honey bee. This suggests that the corpora allata may represent glands, vestigial in the honey bee and many other insects, but functional in at least some of the Lepidoptera. Further research in this direction is much needed.

F. Degenerating Cells

A phenomenon apparently peculiar to the bee, since it is not mentioned by the investigators of other insects, is the frequent occurrence of degenerating cells within the embryonic nerve tissue. These cells occur isolated and in small number within the ventral cord but in the brain they are abundant and to a certain extent localized in definite regions. The largest, most conspicuous and most constant of these include a pair of wedge-shaped sections of the brain, one on each side, situated near the juncture between the proto- and deutocerebrum, and including a part of

[17] *L. c.*, p. 97.

each (Fig. 55, *DegCl*).   Degenerating cells become evident in
these regions as early as Stage IX, and are easily recognizable
at low magnifications by the presence among the brain cells of
deeply stained granules.   Under a high magnification these granules prove to be the remains of the chromatic contents of the
nuclei of degenerating cells (Fig. 62).   Such nuclei are spherical

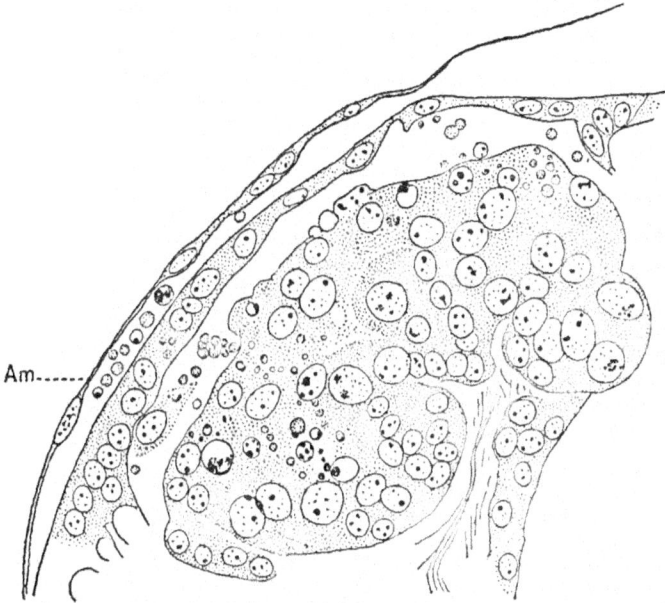

FIG. 62.   Part of the section represented by fig. 56, showing degenerating cells in the protocerebrum, x 600.

in form and smaller than the adjacent nuclei, the nuclear membrane often faint.   The chromatin generally appears to be condensed into one or more relatively large granules, and large and
deeply-stained spherical nucleoli are also commonly present.
Janet (1907) has observed very similar nuclei in the degenerating
wing muscles of an ant (*Lasius*).   The cytoplasm of the degenerating cells becomes broken up into a number of minute spherules,
producing an appearance suggestive of an emulsion (Fig. 62).
This process of cell degeneration continues up to the time of
hatching, and probably even later, but the number of degenerating
cells present in the brain reaches its maximum at Stages XIII
and XIV (Figs. 56 and 62).   At this time not only do the regions
referred to appear to be crammed with degenerating cells, but a
large number of the superficial cells of the protocerebral lobes,
as far back as the optic lobes, are also in a state of degeneration.

A considerable number of the degenerating cells are not retained within the brain tissue; as early as Stage X the debris
of degenerating cells, consisting of the shrunken nuclei and the
spherules just described, may be seen on the dorso-lateral surface
of the brain, and this debris remains here beneath the amnion
apparently unaltered up to the time of hatching, as figure 62
shows. Degenerating cells continue to be extruded from the brain
up to Stage XV and possibly even later. At Stages XIII and
XIV, when the maximum number of degenerating cells are found
in the brain, the characteristic debris of these cells is visible within
the space between the hypodermis and the brain in almost every
section, and is especially abundant over the dorso-lateral region
between the proto- and deutocerebrum, where degenerating cells
are most abundant (Fig. 62).

At Stage XV, that is, just subsequent to the hatching of the
larva, the number of degenerating cells has greatly diminished,
these having either been expelled or absorbed by the brain. A
few, however, are still visible. The debris lying on the outside
of the brain, within the head capsule, has also disappeared, having
probably been washed away by the blood current.

The significance of this extensive cell degeneration is not apparent. That these cells are not artifacts is shown by the uniformity
with which they occur, being present in all series of sections
examined of embryos which were older than Stage IX. Their
presence may accordingly safely be considered as normal. Viallanes (1891), Wheeler (1893) and Heymons (1895) state that
the neuroblasts, or at least the major part of them, degenerate at
the close of embryonic development—a kind of senile degeneration
of the cell. This, as already stated, is not the case in the honey
bee. Moreover the degenerating cells begin to appear at a period
when the neuroblasts are just beginning to be differentiated. The
possibility that some of the neuroblasts degenerate is however by
no means excluded. At Stages XIII and XIV, many of the
superficial cells on the dorsal side of the protocerebrum are in a
state of degeneration, but these are in most cases at least, peri-
pheread of the neuroblasts. This seems to indicate that these
degenerating cells are derivatives of the neuroblasts, which have
wandered peripherad of the latter, as do the neurilemma cells
(p. 160), but their precise origin is unknown.

IX

TRACHEAL SYSTEM, ENDOSKELETON AND HYPODERMIS

1. *Tracheal System*

The essential features of both the structure and development of the tracheal system have been correctly described by both Bütschli (1870) and Grassi (1884). The tracheal system of the larva consists of a pair of longitudinal trunks (Figs. 63, 64 and 75, *TraTr,* Fig. XV, *LTraT*), each of which traverses virtually the entire length of the trunk close beneath the lateral hypodermis. These trunks are connected to the ten spiracles (*Sp*) of each side by slender short branches. At their anterior and posterior ends, respectively, the two lateral trunks are united by semicircular loops or commissures, one of which is situated cephalad of the mid-intestine, above the oesophagus (Figs. 63 and 64, *ATraL,* Fig. XV), the other caudad of the mid-intestine, just ventrad of its juncture with the hind-intestine (Fig. 63 *PTraL,* Fig. XV). The tracheal trunks are also united between these points by eleven other commissures, segmentally arranged. These commissures arise from the ventral side of the lateral trunks, nearly opposite the point of origin of the branches to the stigmata, thence they pass downward, close to the hypodermis and beneath the ventral cord (Figs. 63 and 64, *TraCom,* Fig. XV). The first and second of these loops join the main trunk close to the anterior boundary of the second and third trunk segments, and intersect the ventral mid-line between the first and second trunk segments. The third and fourth loops are united at their bases, joining the main trunk in the anterior half of the fourth trunk segment. These loops intersect the ventral mid-line near the posterior margin of the third and fourth trunk segments. The remaining seven loops are contained entirely within the limits of the corresponding segments. Just in front of each of the spiracles the tracheal trunk sends off one or more branches which pass dorsad, breaking up into finer branches supplying the dorsal region of the trunk. Small

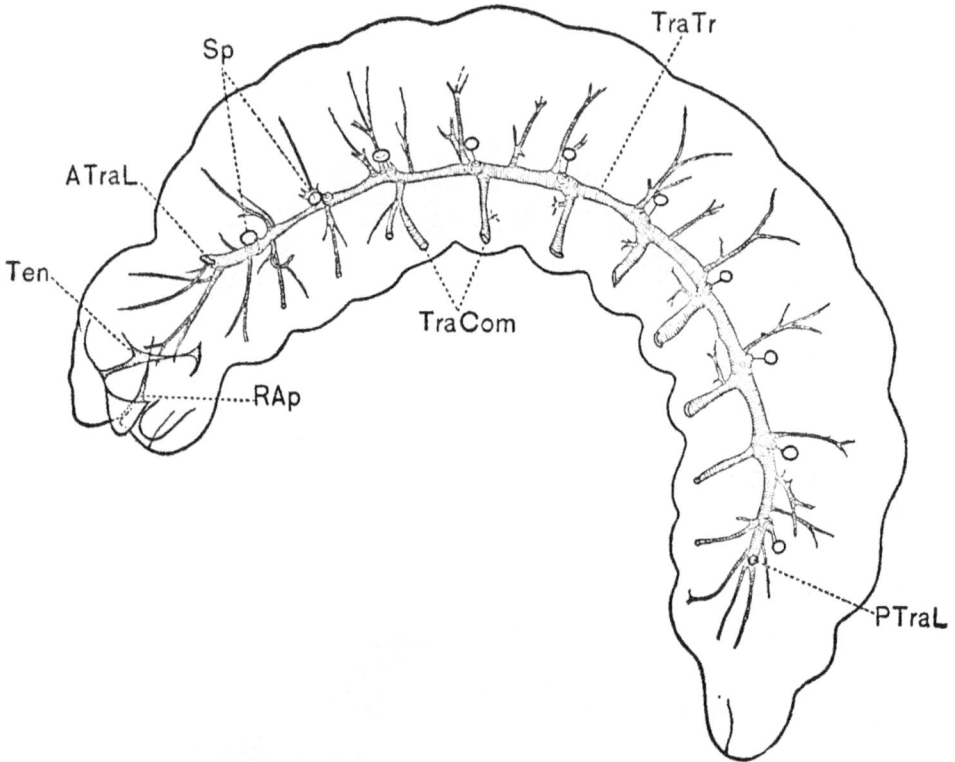

FIG. 63. Side view of larva about two and one-half days old, treated with caustic potash, showing tracheal system.

branches are also sent off by the ventral commissures. At the anterior end of each of the lateral trunks a large branch arises which passes cephalad and breaks up into two smaller branches, both of which run cephalad, one in contact with the inner surface of the optic lobe, the other parallel to the mandibular apodeme and close to it. Mesiad of this large branch is a smaller branch which is sent off by the anterior tracheal commissure. This also extends cephalad supplying the dorsal part of the head. A third branch arises from the ventral side of each of the lateral trunks just cephalad of the first pair of stigmata and dividing, supplies both the base of the labrum and the ventral portion of the first trunk (thoracic) segment (Fig. 64).

The tracheae are simple in structure, being merely tubes composed of a single layer of epithelial cells continuous with those of the hypodermis at the spiracles and lined by chitin thickened in the form of fine transverse threads, more or less spirally arranged, the *taenidia* (Fig. 65, *Tae*).

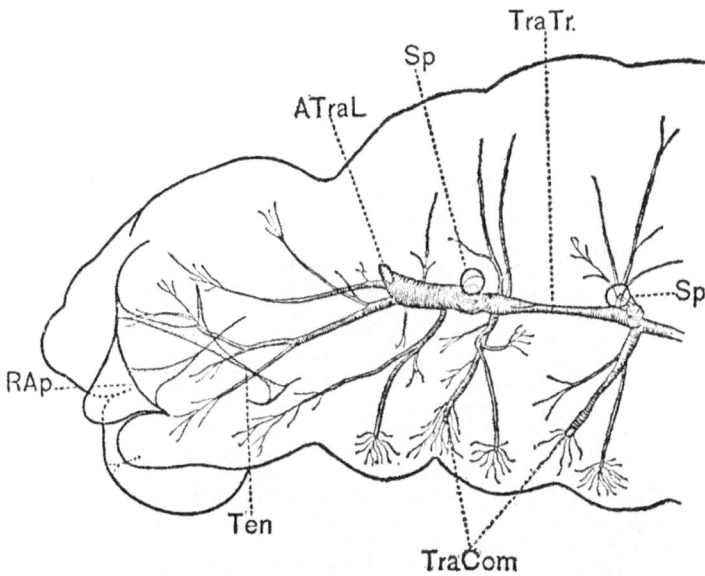

FIG. 64.   Side view of anterior end of larva about two and one-half days old, treated with caustic potash, showing the tracheae and the endoskelleton of the head.

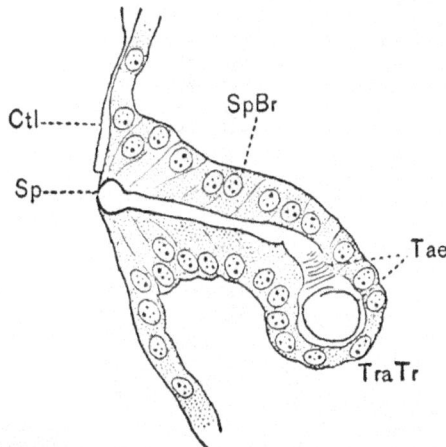

FIG. 65.   Section through a spiracle (Sp), spiracular branch (SpBr) and tracheal trunk (TraTr), of a newly hatched larva (Stage XV), x 600.

The spiracles are twenty in number, ten on each side, ranged in a row along the lateral surface of the trunk, and correspond to the second to the eleventh trunk segments, inclusive. All of the spiracles are situated in the anterior half of the segment to which they belong, the first three pairs lying close to the anterior margin of the corresponding segments. The form and structure of a spiracle, together with the branch connecting it with the main trunk is illustrated by figure 65, taken from a larva just hatched. The spiracle is seen to be minute, in fact scarcely perceptible. In older larvae it is relatively much larger. This opening leads into a spherical chamber, as shown in the illustration, and this in turn into a contracted passageway, the spiracular branch (*SpBr*), which widens out just before reaching the main trunk (*TraTr*). The taenidia (*Tae*) are evident in the main trunk and at the inner end of the connecting branch, but are replaced by a smooth chitinous cuticle for the remainder of the course. This chitin becomes thickened as it approaches the spherical chamber and is continuous with the cuticle of the body at the external opening. Surrounding the cavity leading inward from the spiracle is a single layer of long prismatic cells, continuous with those of the adjacent hypodermis, and forming a conical mass which may be considered as the stigma, and which passes imperceptibly into the cellular portion of the connecting branch, the cells becoming progressively shorter up to the junction with the main trunk.

The rudiments of the tracheal system make their appearance at Stage VIII (Figs. VIII, VIIIa) as two rows of pit-like invaginations of the ectoderm, each row being situated about half way between the lateral edges of the germ band and the ventral mid-line. There are on the trunk ten pairs of these invaginations, corresponding to the second to the eleventh trunk segments. The two or three most anterior pairs appear first, the remainder in rapid succession. The mouths of these invaginations are at first irregular, the first pair being slightly larger than the others. At Stage IX (Fig. IX) these have acquired the form of transverse slits, the first pair however being turned obliquely to the long axis of the embryo. Shortly after this—at Stage X (Fig. X)— they have contracted to minute circular apertures which persist as the spiracles. The invaginations themselves, at first shallow depressions of the ectoderm, increase rapidly in extent and at Stage IX assume the form of flat sacs wedged in between the mesoderm and ectoderm (Fig. 67, *TrInv*). At Stage X, each of these sacs, with the exception of those of the second and eleventh trunk segments, sends out four outgrowths or diverticula, one of which is anterior, one posterior, one dorsal, and one ventral. The anterior and posterior outgrowths of the sacs of the same side meet and fuse with one another to constitute the longitudinal trunk (Fig. 66, *TraTr*), while the ventral outgrowth (*TraCom*) of each sac continues to extend ventrad until it meets with the corresponding one of the opposite side. These then fuse to form the transverse commissures (compare Figs. XI and XII). The dorsal outgrowths continue to extend in the same direction, at the same time breaking up into ramifying twigs, and form the dorsal branches (Fig. 66). The anterior outgrowths

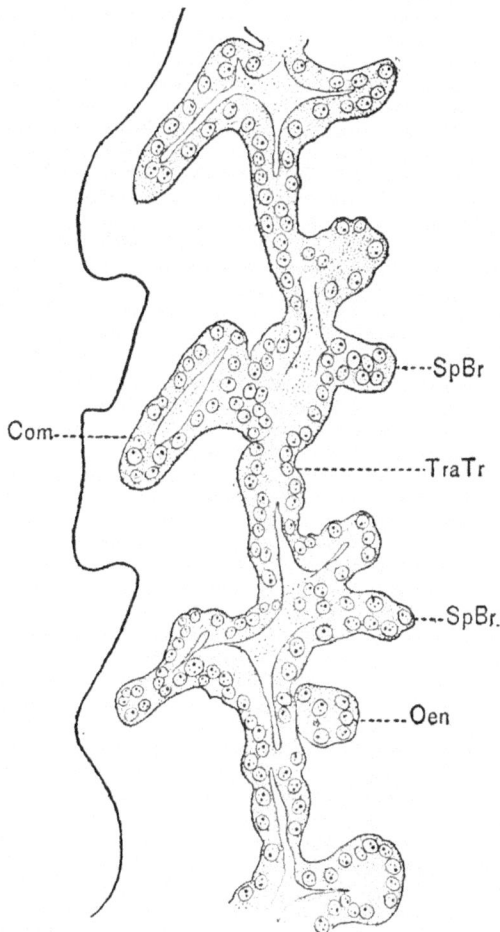

FIG. 66. Sagittal section through the rudiments of the tracheal system of one side in trunk segments 2-4, of an embryo Stage X-XI. Developing oenocytes (*Oen*) also shown, x 387.

of the first pair of tracheal sacs take a course cephalad and dorsad over the mid-intestine and meet and fuse at the dorsal mid-line to form the anterior tracheal loop. The posterior outgrowths of the last pair of tracheal sacs similarly unite and fuse with each other beneath the hind-intestine to form the posterior tracheal loop.

The outgrowths forming the various parts of the tracheal system are at first stout and thick-walled, but as they lengthen they decrease rapidly in diameter until they attain their final form and distribution.

The chitinous lining of the tracheae is formed between Stages XIII and XIV.

In addition to the ten pairs of tracheal invaginations found in the trunk, a pair of tracheal invaginations occur also on the first maxillary segment. These make their appearance at the same time—Stage VIII—as the other tracheal invaginations, and are situated on the lateral surface of the anterior half of the segment, above the base of the rudiment of the second maxilla of each side, and only a few sections caudad of the boundary between the first and second maxillary segments. The location on the segment of this pair of tracheal invaginations is therefore the same as that of the remaining ten. Like the other tracheal invaginations, those of the second maxillary segment are at first shallow and cup-shaped depressions. They soon lose this form and, developing with surprising rapidity, become sacs with a narrow mouth, directed obliquely caudad. The bottom of the sac quickly spreads between the ectoderm and yolk (Fig. 67, *TrInv*) and at the same time sends off four branches or diverticula. One of these is directed caudad, one dorsad and the other two cephalad. Of these last two branches one passes above the mouth of the invagination and one below it. All of these changes take place between Stages VIII and IX. Even before the close of Stage IX the posterior branch of each of these tracheal sacs may be traced back to a juncture with the anterior prolongation of the tracheal sac of the corresponding side of the second trunk (thoracic) segment. The anterior ends of the tracheal trunks therefore owe their origin to the tracheal invaginations of the second maxillary segment. The mouths of the invaginations have by this time contracted to narrow ducts while over them (Fig. 67) on each side is seen a well marked fold of ectoderm obviously caused by the formation of the tracheal sac, which raises the overlying ectoderm, on account of the resistance offered by the yolk. This fold is

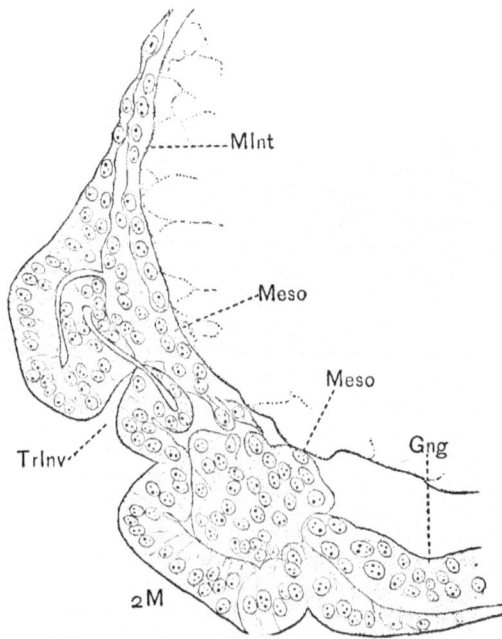

Fig. 67. Transverse section through the anterior half of the second maxillary segment of an embryo, Stage X, showing one of the pair of tracheal invaginations (*TrInv*) and the mesoderm (*Meso*) belonging to this segment, x 387.

clearly shown in figure Xa, but not labeled. At the beginning of Stage X the opening of the tracheal sac to the exterior has already become contracted, and by the end of this stage or the beginning of the next (XI) is completely closed, leaving behind no trace of its existence. At the same time the ectodermal folds also disappear, their disappearance being in part at least due to the shrinkage and withdrawal of the yolk from this region.

The further history of the three remaining diverticula of each of the tracheal sacs remains to be considered. All of these elongate rapidly, assuming a tubular form. The dorsal diverticulum, which is the largest of the three (Fig. 68, *ATraL*), continues to extend dorsad, taking a slightly cephalad course, skirting the posterior margin of the cerebral lobes, along the line of junction of the head and trunk. At Stage XIII the distal ends of these branches reach the dorsal mid-line, when they unite to form the anterior tracheal loop or commissure (Figs. 63 and 64, *ATraL*). Of the two remaining diverticula the ventralmost forms the larger

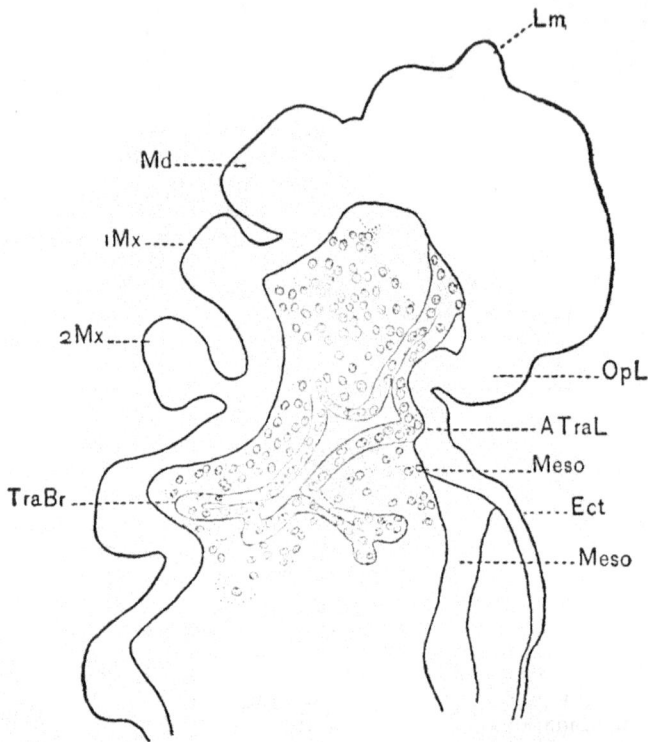

Fɪɢ. 68.   Anterior end of a sagittal section laterad of the median plane, from an embryo of Stage XI-XII, showing the development of the derivatives of the tracheal invaginations of the second maxillary segment. Hypodermis and brain represented in outline only.   Drawn from two sections, x 243.

of the two branches previously described (p. 168), supplying the brain and first maxillae, the other forms the smaller branch supplying the dorsal part of the head (Fig. 64).

Hatschek (1877) described what he thought were tracheal invaginations on the gnathal segments of embryos of *Bombyx,* but examinations of Hatschek's plates show that these invaginations are merely those of the tentorium and mandibular apodemes, and this is the interpretation of subsequent embryologists.   With this exception there is no record in the literature of the insect embryology of tracheal rudiments existing in the gnathal region.   It may therefore be inferred either that this phenomenon is peculiar to the honey bee, or else that it is of more or less general occurrence, but up to the present time has been overlooked.   Of these two inferences the latter seems the more probable.   If it were not so, it would be difficult to understand why the honey bee, a specialized member of a group generally regarded as highly specialized, alone displays a developmental feature highly generalized in its nature, since it can hardly be regarded as a new formation. Moreover the rapidity of the development of these tracheal sacs and the brief duration of their external openings make it still more probable that they have escaped the notice of the investigators of the embryology of other insects. Finally, it is a trait of human nature to see only what is looked for, in proof of which it may be said that the discovery of the tracheal invaginations of the second maxillary segment in the honey bee was scarcely more than a happy accident.

The presence of a pair of tracheal invaginations in the second maxillary segment assists materially to attain a correct conception of the significance of the other invaginations of the cephalic ectoderm. Palmen (1877), Hatschek (1877), Wheeler (1889) and particularly Carrière and Bürger (1897) contended for the homology of the silk glands, the tentorial invaginations and the mandibular apodemes with tracheae. According to Carrière and Bürger the first pair of tentorial invaginations were to be considered as the first pair of stigmata, the mandibular apodemes the second, the second pair of tentorial invaginations the third, etc. Now that a pair of tracheal sacs is known to exist in the second maxillary segment, the homology of the second pair of tentorial invaginations with the stigmata of the second maxillary segment is completely excluded, and the homology of similar invaginations with those of the tracheae is made decidely problematical, especially when one considers that the tracheal invaginations of the second maxillary segment are situated some distance dorsad to the bases of the appendages of that segment. The basis for the conjecture of Janet (1899a, 1900) that the corpora allata furnish material for the tracheae of the head, is likewise removed.

Aside from the occurrence of a tracheal invagination in the gnathal region, the development of the   cheal s . em of the honey bee possesses no points of speci'l teresi. at conforms to the type found in the majority of insects, and th efore requires no further comment.

2.   *Tentorium and Mandibular Apodemes.*

The larval tentorium (Fig. 69B) consists of a narrow chitinous bar or central body, compressed in a dorso-ventral direction,

placed transverse to the long axis of the body and situated
between the oesophagus and the suboesophageal ganglion (Fig.

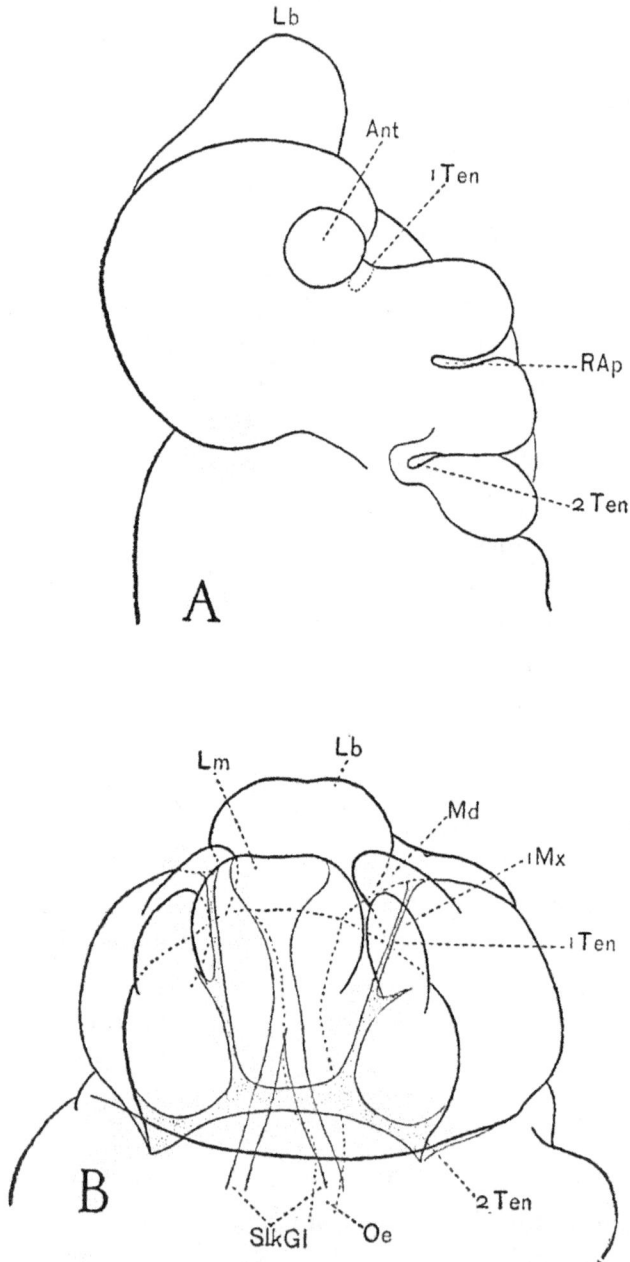

Fig. 69. A, side view of the head of an embryo, Stage XI, drawn in
outline, showing two of the four invaginations which form the tentorium
(*1Ten, 2Ten*), and also the one of two invaginations which form the
mandibular apodemes (*RAp*), x 243. B, ventral view of the head of a
larva about three days old, treated with caustic potash, showing the form
and relations of the tentorium (*1Ten, 2Ten*), x 200.

45 *Ten*). From the ends of this bar spring two pairs of arms, one pair anterior (*1Ten*) and the other posterior (*2Ten*). These arms join the chitinous cuticle of the head. The tentorium thus serves as a set of braces to support and strengthen the cranial capsule as well as to serve for the attachment of muscles. The posterior arms (*2Ten*) together with the central body (*Ten*) form a flat arch, its convex side directed cephalad. The outer ends of these arms are attached to the cranial capsule low down on its lateral aspect, near its junction with the trunk. The anterior arms, which are much more slender than the posterior, arise also from the ends of the central body and are here about equidistant both from one another and from the lateral wall of the head. Each takes a course dorsad, and only slightly laterad, to a point on the anterior cranial wall just above the base of each mandible. About midway between the central body and the anterior cranial wall the anterior arms give off a short pointed spur, directed laterad and cephalad (Fig. 69B). In all of the more advanced larvae studied the anterior arms are much longer than the posterior, but in recently hatched larvae (Fig. XV) the arms are nearly equal in length. As is evident in figure XV the anterior arms together with the central body form a U-shaped structure which embraces the oesophagus and the circumoesophageal commissure.

The tentorium serves for the attachment of two pairs of large muscles, besides a group of small muscles near the mid-line. Both pairs of large muscles are inserted on the wall of the head. The first of these is attached to the tentorium at the ends of the central body, and from here each member of the pair runs ventrad, ectad and caudad, to an attachment of the hypodermis at the base of the first maxilla. The muscles of the second pair are attached to the anterior arms of the tentorium, just cephalad of their junction with the central body, and run obliquely dorsad, laterad and cephalad to a broad area of attachment on the cranial wall, laterad of the cephalic lobes. It is presumably the tendon of this muscle which forms the spur on the anterior arms (Fig. 69B). Muscle fibres also pass from the anterior margin of the central body to the posterior (ventral) wall of the pharynx and doubtless serve as dilators of the latter.

In insect anatomy the tentorium, as well as other portions of the endoskeleton, is commonly regarded as a structure which is wholly chitinous. In sections through the head of a larval bee however the tentorium appears as a system of tubes, composed of a single layer of epithelial cells, which are continuous with those of the hypodermis at the four points where the tentorium joins the head capsule (Figs. 41 and 42,*Ten*, Fig. 62, *2Ten*). The chitinous lining of these tubes constitutes the tentorium proper. This lining is a mere pellicle in recently hatched larvae (Stage XV), but somewhat thicker in older larvae, but never much thicker than the larval cuticle with which it is continuous, and consequently it never attains any considerable degree of rigidity.

The mandibular apodeme, the second member of the endoskeleton of the head, is a slender spine, arising near the base of each mandible at its mesial margin and directed cephalad and dorsad (Figs. 63 and 64, *RAp*). In recently hatched larvae (Fig. XV) the mandibular apodeme arises somewhat cauded of the base of the mandible. Like the tentorium, the mandibular apodeme is hollow and consists of an epithelial tube lined with chitin (Fig. 41, *RAp*). The flexor muscle of the mandible, of which the mandibular apodeme is only the tendon, is a large fan-shaped muscle having a broad origin on the dorso-lateral wall of the cranial capsule, near its junction with the neck and just behind the optic lobes (Fig. 42, *RMcl*). This muscle is more evident in recently hatched than in older larvae.

The origin of the tentorium and the mandibular apodeme is simple. At Stage XI, or a little earlier, three pairs of rounded or oval ectodermal invaginations may be observed at the bases of the mouth parts (Fig. 69A). The first pair (*1Ten*) is situated just in front of the bases of the mandibles close to the antennae, the second (Fig. 61A and B, Fig. 69, *RAp*) between the bases of the mandibles and first maxillae. The third (Fig. 69, *2Ten*) between the bases of the first and second maxillae. These invaginations deepen and grow inward as slender tubes, blind at their inner ends (Fig. 61C, *RAp*). The first pair becomes directed caudad, the second caudad and dorsad, the third cephalad and mesiad. At Stage XII the first and second pair meet and coalesce to form the tranverse bar or central body of the tentorium.   The second pair constitute the mandibular apodeme.

Heider (1889), Heymons (1895), Carrière and Bürger (1897), Toyama (1900) and Riley (1904) have given complete accounts of the development of the endoskeleton of the head, and with these the account just given is in substantial agreement, and may possibly be considered therefore as representative of the typical mode of development of this structure.

The tentorium of the honey bee is unmentioned in all previous accounts except that of Grassi (1884).   In this are described and figured the two pairs of invaginations which ultimately unite to form the tentorium, but in later stages Grassi lost sight of the anterior pair and consequently failed to grasp their significance, supposing that only the posterior pair were concerned in the formation of the tentorium.

3.   *Hypodermis*

The hypodermis is formed of the ectoderm which remains after the subtraction of the various tissues and organs of ectodermal origin.   At Stages VII and VIII, the dorsal half of the blastoderm is stripped from the yolk to form the amnion.   This portion of the yolk is therefore left entirely bare.   A little later—at Stage IX— a delicate cellular membrane is found covering this previously nude area, except in the cephalic region, which now is covered by the cephalic end of the germ band.   This membrane is as thin and delicate as the amnion itself, and is continuous laterally with the ectoderm, which therefore is presumably responsible for its production.   This thin dorsal sheet, for the sake of convenience in description, may be termed the *extra-embryonic* ectoderm, in contradistinction to the *embryonic ectoderm,* which overlies the mesoderm, and from which are derived the oenocytes, the tracheal system and the ventral nerve cord.   The subtraction of this material from the embryonic ectoderm, which takes place during Stages IX-XI Figs. 78 and 79), materially reduces the thickness of the embryonic ectoderm, particularly in its median part, from which the ventral cord is derived.   At Stage XI (Fig. 80) the entire embryo is seen to have begun to increase in breadth, this

increase naturally affecting the embryonic ectoderm, which always
remains slightly wider than the underlying mesodermal tissues.
During the two succeeding stages, XII and XIII (Figs. 81 and
82), the embryo continues to increase in breadth, its lateral limits
being marked by the lateral—now dorsal—margins of the embryonic ectoderm which is progressing upward over the sides
of the egg toward the dorsal mid-line. This growth of the embryonic ectoderm, as the figures indicate, takes place in part at
the expense of the extra-embryonic ectoderm, which, with the
increasing breadth of the embryonic ectoderm, undergoes a corresponding decrease. Meanwhile the embryonic ectoderm is
steadily diminishing in thickness, and at its margin it passes
rather gradually into the thinner extra-embryonic part. Between
Stages XIII and XIV, the lateral margins of the embryo meet
and unite along the dorsal mid-line, the extra-embryonic ectoderm
being now entirely absorbed by the embryonic ectoderm, now the
definitive hypodermis (Fig. 82).

The head capsule is formed as follows: During the interval
between Stages VII and VIII, when the cephalic end of the germ
band, consisting principally of the protocerebral lobes, travels
around the cephalic end of the egg, the interval between the
lateral margins of the procephalic lobes and of the adjoining
margins of the gnathal segments remains continuously filled by
ectoderm. The interval between the procephalic lobes is, at
Stage VII, filled by a thick layer of ectoderm (Fig. 29, *Ect*), and
after the procephalic lobes reach the dorsal side of the egg,
this part of the ectoderm constitutes the dorsal part of the head
capsule, comprising the triangular area between the cerebral
lobes. The remainder of the head capsule is formed from the
superficial layer of that part of the cephalic ectoderm concerned
in the formation of the brain.

X

The Oenocytes

The name *oenocytes* was given by Wielowiejsky (1886) to certain large wine-colored cells observed embedded in the fat body of the larvae of *Corethra,* and which he also found in other insects belonging to several different orders.

This investigator was, however, not the first to note these cells, for as Graber (1891) has stated, similar cells had been previously observed and described by other investigators, as for example, by Tichomiroff (1882), in the embryos of the silk worm. After Wielowiejsky called attention to the oenocytes they were observed in the larvae and imagoes of insects belonging to so many different orders that they are now generally assumed to be common to all pterygote insects. They differ much in different orders; the principal features which they have in common appear to be their relatively large size and their ectodermal origin. To this it may be added that after leaving the ectoderm they have never been seen to divide. They are always found more or less closely associated with the cells of the fat body, and often attached to tracheae. They may be scattered or arranged in clusters, in which case a pair of clusters is usually situated in each of the first eight abdominal segments, as in *Corethra.* Little is known of the functions of the oenocytes, although several attempts have been made to solve this problem, the latest being that of Glaser (1912), who has studied the oenocytes of the larvae of the leopard moth and finds evidence that they contain oxydising enzymes.

The development of the oenocytes has been observed by several investigators. Graber (1891) calls attention to the fact, that prior to the appearance of Wielowiejsky's paper, Tichomiroff (1882) and Korotneff (1885) had described the development of cells to be regarded as identical with oenocytes, and that later Heider (1889) had also observed an ectodermal proliferation which was probably concerned in the development of oenocytes, but without recognizing its significance. Graber also described the develop-

ment of the oenocytes of the grasshopper *Stenobothrus*.   Here
the oenocytes are derived by delamination from the inner surface
of the lateral ectoderm.   In the beetles *Hydrophilus, Melolontha*
and *Lina* Graber finds presumptive evidence that the oenocytes
are derived from the ectoderm near the eight abdominal stigmata,
including special "metastigmatic" invaginations, one of which is
situated directly caudad to each of the eight abdominal stigmata.[18]
In *Hydrophilus* and *Melolontha* these invaginations are situated
near the posterior border of the segments some distance behind
the spiracles, in *Lina* they are close behind them.

Wheeler (1892) has reviewed the work of previous investigators and described briefly his observations on the development
of the oenocytes in the representatives of the Orthoptera, Ephe-
meridea, Hemiptera, Neuroptera and Lepidoptera.   In all cases
they were found to arise by either delamination or migration from
the lateral ectoderm, and the metastigmatic invaginations of
Graber were found not only in representaives of the Coleoptera
but also in representatives of the Orthoptera (*Blatta, Xiphidium*),
but Wheeler believes that they play only a minor part in the production of the oenocytes.

Heymons (1895) studied the development of oenocytes in
Forficula and *Gryllus*.   He states "that by means of simple
migration or through the intervention of Graber's metastigmatic
invaginations they find their way into the interior of the
body and thence into the already formed fat body.   Here for a
long time they form sharply marked, metamerically arranged
groups of cells.   Later they become more irregularly distributed
throughout the abdomen, and also find their way into the thorax,
but are nevertheless to be here easily distinguished from the true
fat cells principally by their size."   So far this corresponds
closely with Wheeler's account (1892) of *Blatta* and *Xiphidium*.
Heymons was, however, able to add the important fact that
oenocytes are developed from all of the eleven abdominal segments, and not merely from those bearing spiracles, as had been
previously assumed to be the case.   Moreover these "postspiracu-
lar" oenocytes differ from the remainder by their larger size.

[18] Wheeler (1892) states that Patten (1889) had previously observed
these in *Acilius*.

Hirschler (1909) finds that in *Donacia* the oenocytes arise in two ways. A part of them arise by migration from restricted areas of eight of the abdominal segments, immediately caudad of the stigmata, thus forming metamerically arranged groups. Another part is formed by a sort of diffuse or irregular immigration from various parts of the ectoderm, but particularly from the vicinity of the pericardial septum and the proctodaeum.

The oenocytes of the honey bee were first described by Wielowiejsky (1886) and have been studied in detail by Koschevnikov[19] (1900, 1905), and—in connection with the metamorphosis —by Anglas (1900). These investigators agree that in the larval honey bee the oenocytes are represented by "certain cells which are of great size, not vacuolated, their protoplasm staining strongly with borax carmine, and having large regular oval nuclei. . . . They are to be found in the depths of the body surrounded on all sides by the fat body and sometimes packed close against the walls of the Malpighian tubules. I never observed two nuclei in any of these cells. During histolysis, they float free in the body cavity. The largest of them are truly gigantic, measuring 176 micra with a nucleus 56 micra in diameter."[20] One of these cells, together with the surrounding fat cells, in a larva four days old, is illustrated in figure 70A. In younger and smaller larvae the oenocytes are smaller, but their size, as compared with that of the fat cells, remains about the same. An oenocyte surrounded by three fat cells, from a recently hatched larva, is represented in figure 70B. It will be noted that in this case the magnification is 1107 diameters, while in that of the older larva (Fig. 70A) the magnification is only 580 diameters. A fair conception of the

[19] Glaser (1912) has expressed doubt in regard to identity of the cells described by Koschevnikov with the oenocytes of other insects on the ground of Koschevnikov's statement that he has observed the oenocytes in the act of engulfing fat cells. Without attempting to weigh the merits of this objection, it may be remarked that the oenocytes of Koschevnikov, in the larval stages at least, correspond morphologically with the oenocytes of other insects more closely than any other cells within the body cavity, and, if these are not truly oenocytes, it would be necessary to assume that such cells are either non-existant in the bee, or else that they are morphologically different from those of other insects. It is of interest to note in this connection that Anglas (1900) states that he has never seen them act as phagocytes.

[20] Koschevinkov (1909).

great increase in size of the oenocytes and fat cells between the newly hatched larva and the larva four days old may be gained by comparing figures 70A and 75, *Ocn,* these two figures being drawn at the same magnification.

As in older larvae the oenocytes at Stage XV are principally distinguished by their size which, although somewhat variable, is only equalled by the neuroblasts and some of the blood cells. Their form is largely determined by their location. When free they are more or less spherical as in figure 70B; when in close contact with other cells, polyhedral, as in figure 70A. The nucleus is usually more or less spherical and its chromatin content is in young larvae not especially different from that of the cells of other tissues. The cytoplasm presents a rather more homogeneous appearance than that of the fat cells. The oenocytes are rather irregularly scattered throughout the body cavity and are frequently enmeshed in the fat cells, as in figure 70B. At Stages XIV and XV oenocytes are rare in the first three trunk segments, in other words the oenocytes during these stages are found principally in the last twelve trunk segments, corresponding to the imaginal abdominal segments of insects of other orders than the Hymenoptera.

In surface views of many embryos of Stage X small clear spots, like those of the rudiments of the spiracles, may be seen to alternate with the last eight spiracles of each side, so that the latter have the appearance of being double. Transverse sections show that there are actually at this stage pit-like invaginations of the ectoderm precisely in line with the spiracles and that each of these invaginations is situated half way between the adjacent spiracles.

Each pair of spiracles is, however, situated, not in the middle, but in the anterior half of the corresponding segment, so that each pair of the invaginations just described belongs to the posterior half of the same segment as the pair of spiracles next in front (cephalad of it). Accordingly these invaginations are located on the third to the eleventh trunk segments, or, regarding only the first three trunk segments as belonging to the thorax, the first eight abdominal segments. These invaginations correspond in form, size and location with the "metastigmatic" invagi-

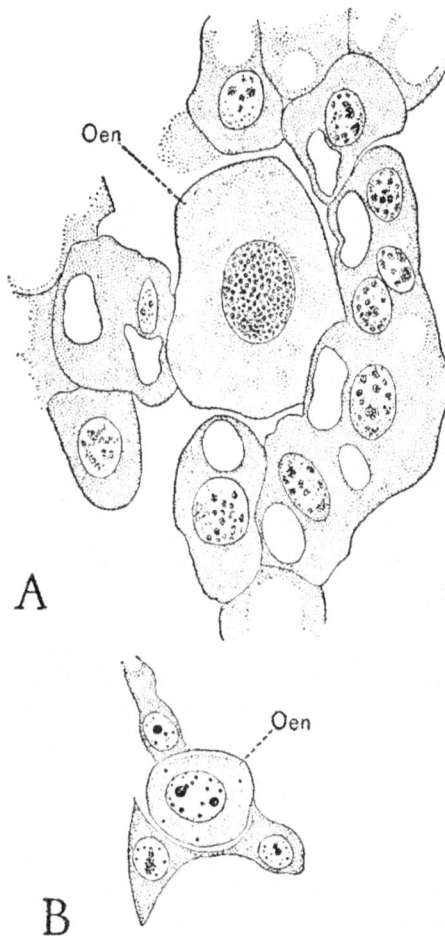

FIG. 70. Oenocytes (*Oen*), surrounded by fat cells. A, from a larva about four days old, x 387; B, from a newly hatched larva, x 1107.

nations of *Melolontha* as described by Graber (1891). Figure 71 shows half of a tranverse section through the region between the fourth and fifth trunk segments and intersecting the first metastigmatic invagination (*Oen*), showing its position and relation to the surrounding tissues. In figure 72A is shown one of these invaginations more highly magnified. There are in all eight pairs of these, as in those insect embryos examined by Graber (1891) and Wheeler (1892). The first pair is the largest, those caudad successively smaller, the last being faint and not easy to discern. Each invagination is funnel-shaped externally. Internally it is formed by a cluster or rosette of long pyriform ectoderm cells as shown in the figure, the entire structure strongly suggest-

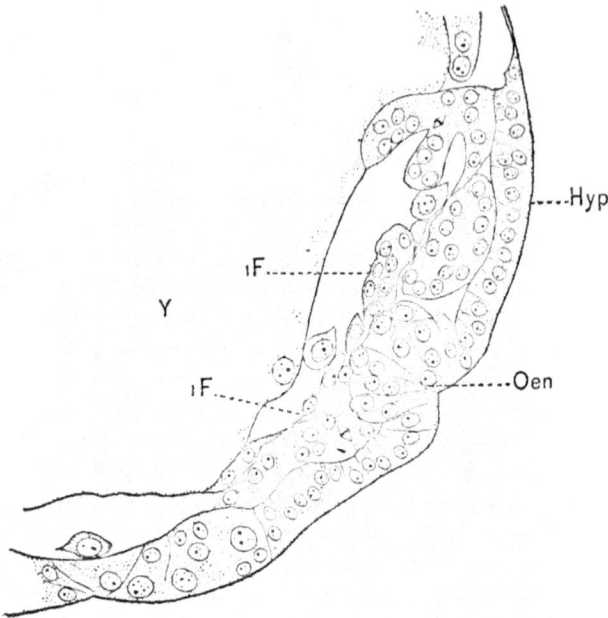

FIG. 71. Right half of a transverse section through the posterior portion of the fourth trunk segment of an embryo. Stage X, showing the location and relative size of the cell-clusters which produce the oenocytes, x 367.

ing a sharp contraction of the ectoderm at this point. Even at this stage some of the cells of the clusters are seen to be detaching themselves and migrating inward. These cells are in all probability the embryonic oenocytes (*Oen*). At Stage XI the external evidences of an invagination have almost disappeared, only a minute depression marking the point where it had existed. The oenocytes are now seen to be slightly larger than the cells of the ectoderm and are moving inward into the body cavity. This is fairly well shown by figure 72B. Certain other sections illustrate this particular point somewhat more clearly, but the section also shows another important but also unfortunate circumstance, that is, the similarity of the oenocytes at this stage to the adjacent mesoderm cells, particularly those of the fat body (*1F, 2F*). This similarity is so close that the continuity of the cells seen at Stage X with the oenocytes as they appear in the late embryonic stages (XIII, XIV), as well as of the larval stages could not be demonstrated, in spite of repeated attempts. Graber (1891) encountered the same difficulty in *Hydrophilus*. Nevertheless, in view of the developmental history of the oenocytes in those forms in

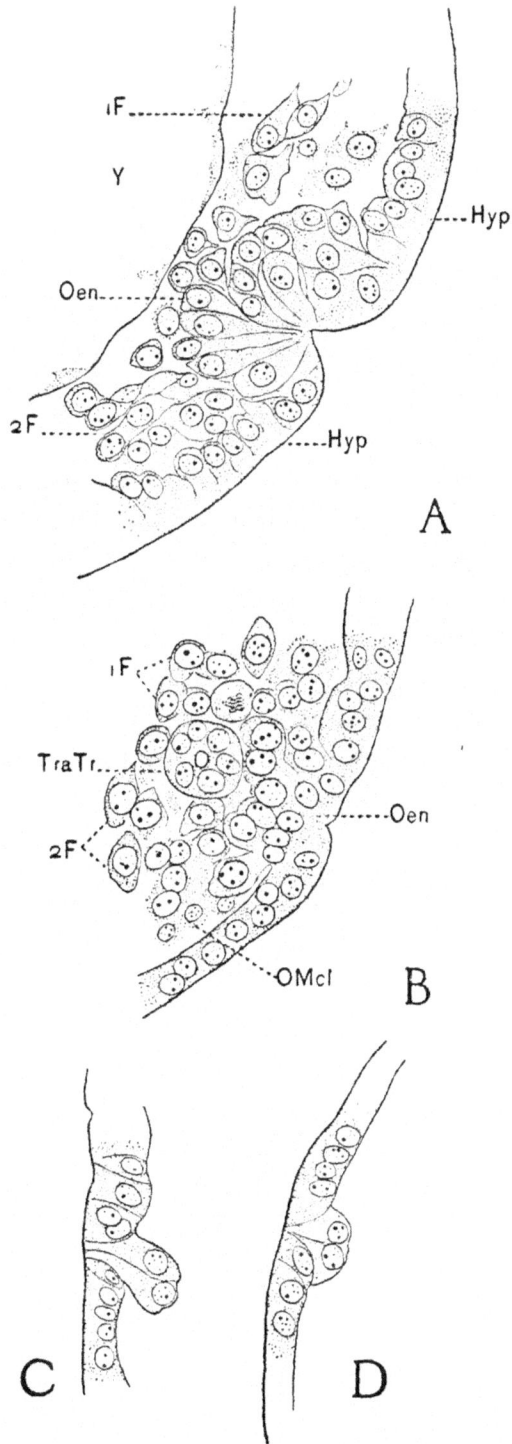

FIG. 72. A, transverse section through the lateral ectoderm of the fourth trunk segment of an embryo, Stage X, showing the formation of oenocytes (*Oen*). B, a similar section from the third trunk segment of an embryo, Stage XI-XII. C, section through the lateral ectoderm of the fourth trunk segment of an embryo, Stage XIII, showing the point where the oenocytes arose. D, the same, from the sixth trunk segment of an embryo, Stage XIII-XIV, x 600.

which it may be continuously followed, as in *Stenobothrus* or *Xiphidium*, one can scarcely doubt that the metameric cell clusters seen in the bee embryo also give rise to the oenocytes of the larva.

At Stages XII-XIV the point of origin of oenocytes is marked by a minute pad or tuft of cells which stain somewhat less deeply than the other hypodermal cells (Figs. 72C and D). In some sections the cells of this tuft are so long as to suggest the possibility of the separation of some of these from the hypodermis (Fig. 72C), but no direct evidence of this was obtained.

The definitive oenocytes of the larva become evident at Stage XIII-XIV, owing to their large size as compared with the other cells in the body cavity. From this time until hatching their size remains practically constant and is the same as that of the cells of the segmentally arranged clusters of Stages X and XI, which are assumed to be the embryonic oenocytes, so that the identification of the oenocytes at the later stages is accordingly made possible by the differentiation of the other cells in the body cavity and the accompanying reduction in their size, rather than by any morphological change in the oenocytes themselves.

XI

MUSCLES, FAT BODY AND CIRCULATORY SYSTEM

In the young larva each muscle is composed of a varying number of parallel fibres, which constitute the contractile portion of the muscle. These fibres range from round to flat in tranverse section, stain deeply and are themselves composed of fine fibrillae (Figs. 73A and B). Each fibre is continuous from its origin to its insertion and remains of nearly the same diameter throughout its length, tapering only slightly toward the two ends. In the longitudinal trunk muscles, however, which are attached to the intersegmental hypodermis, instead of terminating at these points, the fibres appear to be continuous through at least two segments, and probably more (Fig. 73A). A layer of undifferentiated protoplasm, the *sarcoplasm,* enwraps all of the fibres which constitute a single muscle. At the middle of a muscle the sarcoplasm is common to all of the fibres of the muscle, binding them together, but towards the two ends of the muscle the sarcoplasm divides, a small portion, constantly diminishing, following each fibre to its attachment (Fig. 73A). Embedded in the sarcoplasm are the muscle nuclei, arranged in rows along the fibres. Usually they lie all on one side of the fibre—in the case of the longitudinal trunk muscles the inner side—but this rule is subject to many exceptions. In young larvae none of the muscle fibres are transversely striated. In the imago, on the other hand, the greater part of the muscles are striated, while the relation of the sarcoplasm and the muscle fibres is the reverse of that existing in the larva, the sarcoplasm and its nuclei in the imago constituting, so to speak, the core or central portion of the muscle, about which are arranged the muscle fibres. It is of interest to note that the condition existing in the muscles of the bee larva is quite similar to that found in the muscles of the vertebrates.

The muscles may for convenience be divided into three groups: the muscles of the head, the muscles of the trunk, and the muscles of the alimentary canal.

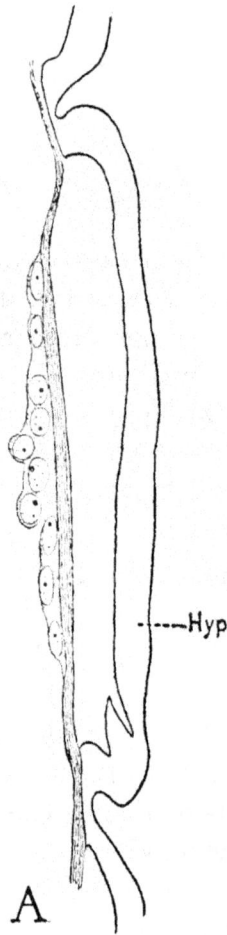

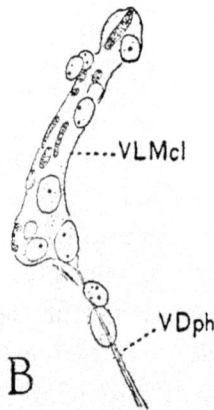

FIG. 73. A, muscle fibre, from a longitudinal section of the dorsal longitudinal trunk muscles, between trunk segments 1 and 2. Hypodermis shown in outline. B, transverse section through one of the ventral longitudinal muscles, x 840.

The muscles included in the head comprise the gnathal muscles, moving the mouth parts, the muscles attached to the tentorium (already described under the section relating to that structure) and certain muscles which although having their origin on the endoskeleton of the head belong properly to the alimentary canal.

The muscles of the newly hatched larva are by no means so easily identified as in older larvae, being less well differentiated. Moreover in so small an object the study of sections is the only method available. In older larvae both mandibles and maxillae are provided with flexor and extensor muscles, those of the mandibles being much the larger. In newly hatched larvae, however, the only gnathal muscles clearly evident are the flexor muscles of the mandibles. These are relatively large muscles, each of which is inserted on the mandibular apodeme, whence it passes dorsad, caudad and ectad, spreading out fanwise meanwhile, to a broad origin on the inner surface of the dorso-lateral face of the head capsule behind the optic lobes and near the junction of the head and trunk (Fig. 42, *RMcl*).

The principal muscles of the trunk comprise dorsal and ventral longitudinal muscles and oblique muscles. The dorsal longitudinal muscles (Fig. 75, *DLMcl*) are broad flat sheets comprising one layer of fibres, and lie on either side of the heart, beneath the dorso-lateral hypodermis extending far down on the side. A pair of these are situated in every segment, their ends being attached to the intersegmental hypodermis, as shown in figure 73A.

The ventral longitudinal muscles (Fig. 75, *VLMcl*) correspond in their arrangement and function to the dorsal longitudinal muscles, but are only about half as wide. They extend laterally from the ventral nerve cord about half way up to the level of the longitudinal tracheal trunk.

The oblique muscles (Fig. 75, *OMcl*) comprise a pair of muscles spanning the lateral wall of each segment and arranged in such a way that if seen in side view they would together form a figure resembling the letter X. The smaller of these lies close to the hypodermis, and runs dorsad and cephalad, while the larger is situated a short distance within and runs ventrad and cephalad. In addition to the oblique muscles there are delicate muscles,

lying close to the inner surface of the hypodermis and consisting
of but three or four fibres, which take their origin on the hypodermis a short distance below the stigmata, on the ventral side
of which they are inserted. These muscles are so insignificant
that they are easily overlooked except in favorable tangential
sections.

The muscles of the alimentary canal comprise a well defined
layer of minute circular fibres enwrapping the whole alimentary
tract, including the fore- and hind-intestines, in addition to which
is a layer of longitudinal fibres, much fewer in number and more
scattered than those of the circular fibres (Fig. 74 *MclEnt*).

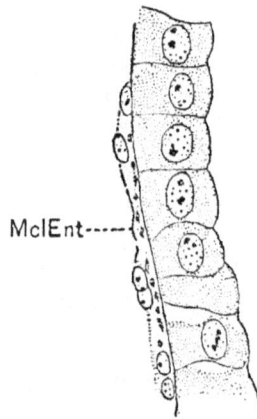

FIG. 74. Part of a sagittal section through the dorsal wall of the mid-
intestine of a newly hatched larva (Stage XV), showing the muscular
coat (*MclEnt*), x 600.

These muscles, both longitudinal and circular, may conveniently
be termed the *enteric muscles*. They are best developed in the
fore-intestine, and here the longitudinal underlie the circular
fibres. In the case of the mid-intestine this relation, as in the
imago, is reversed, the longitudinal fibres lying outside the
circular. This is also true of the hind-intestine. The fibres making up the enteric muscles are extremely delicate, and the muscle
nuclei minute. On the mid-intestine the circular muscle
fibres are fairly numerous and regularly arranged, the longitudinal fibres, on the contrary, are scattered and of such extreme
delicacy that it is only in favorable cases that they can be seen
at all (Fig. 74). Muscle fibres also run from the anterior (dorsal) wall of the labrum to the anterior (dorsal) wall of the

pharynx, while a few fibres run from the anterior margin of the central body of the tentorium to the posterior wall of the pharynx; both of these sets function as dilators of the pharynx.

The muscles associated with the circulatory system will be described under that heading.

The *fat body* of the recently hatched larva has a decidedly different aspect from that in the old larva. In the latter the fat body forms a compact mass filling the spaces of the body cavity and composed of cells loaded with large fat globules, in figure 69A represented by the empty spaces in the cells surrounding the oenocyte. In the young larva, on the other hand, the fat body never constitutes a compact mass, but is made up of a loose network of branching cells (Fig. 69B) in only some of which a minute fat globule can be made out. In the thorax the fat body is somewhat more compact than in the abdomen, in which the fat cells are much scattered, as shown in figure 75, *1F, 2F, 3F*. As in *Chalicodoma* (Carrière and Bürger 1897) the fat body is more or less incompletely divided into three sections: a dorsal section, situated above the dorsal diaphragm, the pericardial fat body (Fig. 75, *3F*), and two sections situated laterad on the mid-intestine (*1F, 2F*) and divided from one another by the longitudinal tracheal trunk.

The *circulatory system* consists of the *heart,* the *dorsal* and *ventral diaphragms,* and the *blood corpuscles.*

The *heart* (Figs. XV, 75, 76, *Ht*) is a slender tube lying in the mid-line between the mid-intestine and the dorsal hypodermis. It extends from the twelfth trunk segment forward to the head where it is continued as the *aorta* between the cerebral lobes, lying immediately above the oesophagus. The anterior end of the aorta opens into the body cavity. The posterior end is closed. The lateral walls of the heart are relatively thick, and contain the nuclei while the ventral and dorsal walls are thin, the latter being in some sections so tenuous as to be scarcely perceptible, the heart having the appearance of being closed dorsad by the hypodermis (Fig. 75). Along the ventral side of the heart is a narrow strip of cells, to which, in trunk segments seven, eight and nine, the ovaries are attached. On its dorsal side the heart is slightly indented by the intersegmental constrictions, while

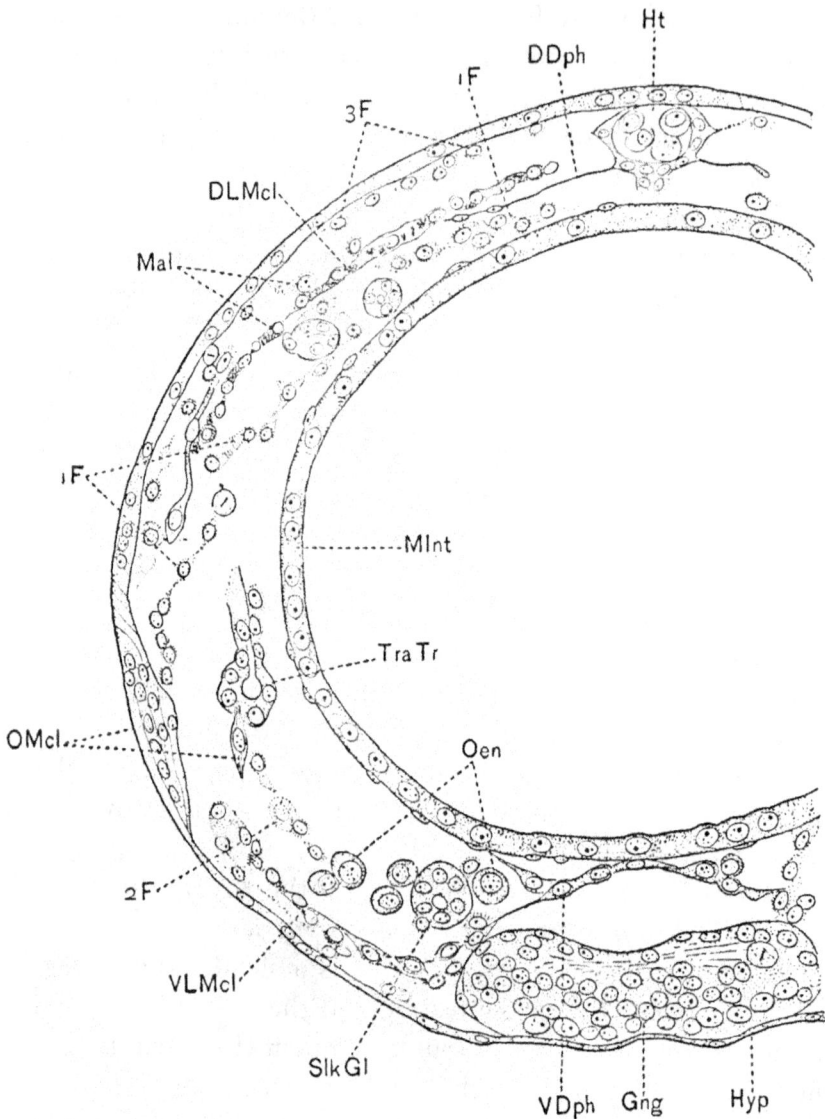

FIG. 75. Transverse section through the 6th trunk segment of a newly hatched larva, x 387.

in the middle of each segment its lateral walls are sharply infolded, reducing the lumen at these points to a narrow vertical slit. These folds evidently correspond to the *ostia*, or lateral valvular openings (see Snodgrass, 1910, p. 107), although no opening in the lateral walls was observed at Stage XV.

In tranverse sections an exceedingly delicate membrane is seen to extend right and left from the ventral wall of the heart, the *dorsal diaphragm* (Figs. 75 and 76, *DDph*). This ends free in the body cavity, except in the intersegmental regions, where it extends laterad to the hypodermis, to which it is attached (Fig. 76,

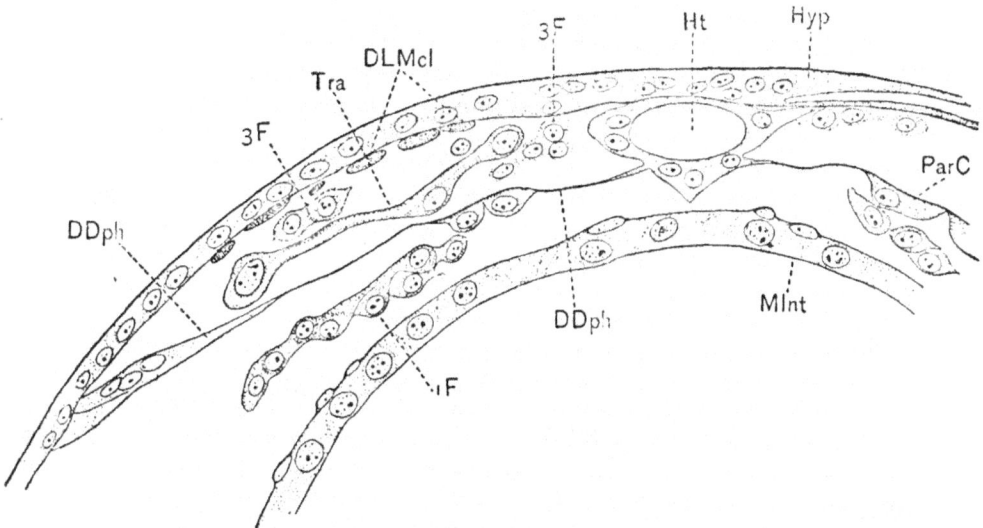

FIG. 76. Transverse section through the dorsal region of the trunk, between the 5th and 6th trunk segments, showing the attachment of the dorsal diaphragm (*DDph*) to the hypodermis, x 600.

DDph). This relation is clearest in the third to the ninth trunk segments inclusive, but undoubtedly also exists in the first two (thoracic), although masked here by the crowded condition of the parts and the large number of pericardial fat cells. The muscle fibres constituting the dorsal diaphragm of most insects, including the honey bee (imago), are commonly described as radiating fanwise from the points of origin to their attachment on the heart (*vide* Snodgrass, 1910, p. 108), but in the young larvae no muscle fibres could be seen in the dorsal diaphragm, which appears in both longitudinal and tranverse sections as a continuous and extremely thin membrane.

The cells constituting the dorsal diaphragm are of two kinds; numerous flat cells with minute nuclei (Fig. 70), appearing to be the diaphragm cells proper, and pale lenticular cells whose size approximates that of the fat cells (Fig. 76 *ParC*). An examination of the text and figures of Carrière and Bürger's (1897) account of the dorsal diaphragm of *Chalicodoma* makes it evident that the cells just described are identical with those constituting the "paracardial cellular cord" (paracardiale Zellstrang). This term was first used by Heymons (1895) to designate segmentally

arranged groups of cells lying laterad to the heart among the
fibres of the dorsal diaphragm of *Forficula.* Although at first
sight it may seem somewhat doubtful whether the scattered cells
of the dorsal diaphragm in the larvae of the honey bee are to be
regarded as the homologues of the segmentally arranged cells
composing the "paracardiale Zellstrang" of *Forficula* and *Chalicodoma,* nevertheless—as will appear later—they have precisely
the same origin in all three insects, so that their homology can
scarcely be open to question. In view of the scattered distribution of these cells in the honey bee it will be more convenient to
refer to them simply as the *paracardial cells.*

The *aorta* into which the anterior end of the heart opens, is
merely the space around the oesophagus bounded externally by
the inner walls of the coelomic sacs of the antennal segment (Fig.
42, *Ao*). It opens cephalad and ventrad into the cavity of the
head.

The *ventral diaphragm* is composed of a layer of transversely
arranged muscle fibres which combine to form a more or less continuous membrane (Fig. 75, *VDph*) overarching a space just
dorsad of the ventral cord, the *ventral sinus.* The ventral diaphragm extends from the first to the neighborhood of the twelfth
trunk segments, its lateral margins being attached to the mesial
margins of the ventral longitudinal muscles and also to the silk
glands. In the mid-line its dorsal surface is in contact with the
ventral wall of the mid-intestine.

The *blood corpuscles* are large rounded cells which at Stage
XV are for the most part confined to the lumen of the heart. In
size they approach the oenocytes, but their general appearance is
quite different. Both nucleus and cytoplasm are pale, the latter
being much vacuolated and frequently enclosing deeply stained
granules. A group of blood corpuscles are shown within the heart
in figure 75.

Since the organs and tissues which form the subject of this
section are all derived from the mesoderm, a consideration of the
development of this germ layer naturally comes next in order.
The development of the mesoderm in the head, including the
gnathal segments, differs from that of the trunk, and since the
latter may be regarded as representing a less modified and therefore more typical condition, it will be considered first.

At Stage VII a tranverse section through the middle of the trunk shows the mesoderm (Fig. 32, *Meso*) as a double layer of cells lying close beneath the ectoderm, between the latter and the yolk. This condition is typical of the mesoderm from the gnathal region at least as far as the end of the eleventh trunk segment. The inner of the two layers is the *visceral*, the outer the *somatic* layer. In the neighborhood of the ventral mid-line the cells of the mesoderm are rather irregular in form and arranged in but a single layer. At the lateral margins of the mesoderm where the two layers are continuous with one another, they are relatively thick, well defined and separated from one another by a narrow slit-like space, while the component cells of both layers are columnar in form. A similar condition obtains in *Chalicodoma,* and to these lateral regions, composed of columnar cells, Carrière and Bürger (1897) applied the name "mesodermal tubes" (Mesoderm-röhre) as descriptive of their form and since these sections of the mesoderm are marked off rather sharply from the remainder both in their form and their behavior, this term will for convenience be adopted in the following description. Between the mesodermal tubes and the single-layered median strip the mesoderm cells are flattened and somewhat irregular in form, particularly those of the visceral layer. In *Chalicodoma,* Bürger (1897) found that the mesodermal tubes were divided intersegmentally by thin partitions, moreover, the mesoderm mesiad of the mesodermal tubes is divided into pairs of flat sacs (mesodermal sacs), a pair to each segment, and communicating only intrasegmentally with the mesodermal tubes of the corresponding side. No well marked evidence of segmentation of the mesoderm could be found in the honey bee. The two layers of the mesoderm on each side of the mid-line are virtually continuous from the second maxillary segment to the eleventh trunk segment, although a careful examination of sagittal sections of Stages VIII and IX seems to indicate that the mesoderm mesiad of the mesodermal tubes is slightly constricted intersegmentally. At Stage VIII-IX (Fig. 77, *Meso*) while the mesoderm has changed but little in its general appearance, it may be noted that the cells in the immediate vicinity of the mid-line have increased in size, and assumed a rounded form. These (*BlC*) are to form the blood corpuscles.

At Stage IX (Fig. 78) the mesoderm begins to show important

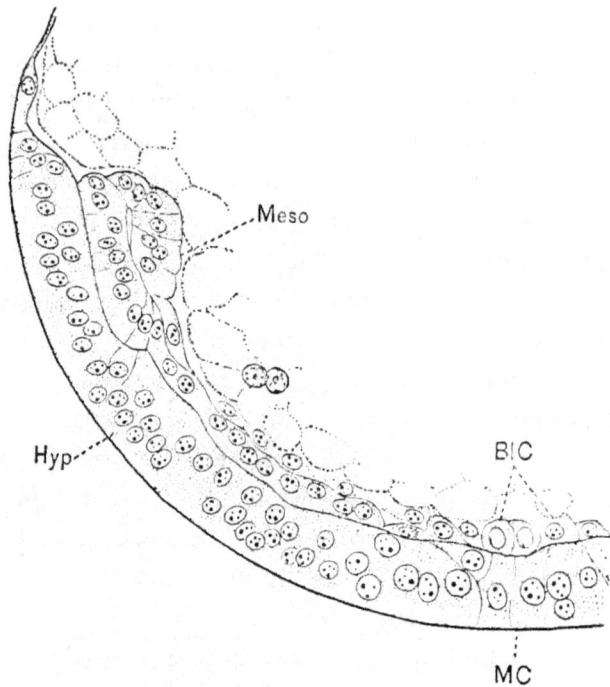

FIG. 77. Transverse section through the future basal region of the abdomen of an embryo, Stage VIII-IX; showing the structure of the mesoderm (*Meso*), x 387.

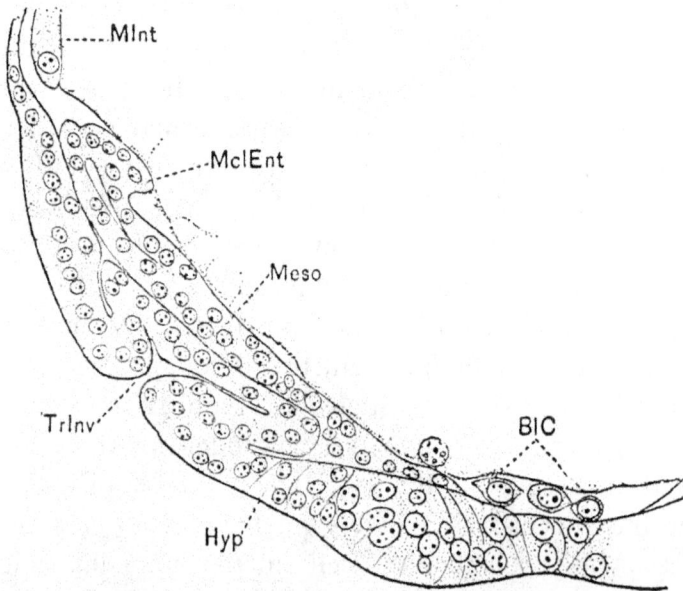

FIG. 78. Transverse section through the fourth trunk segment of an embryo, Stage IX, intersecting one of the tracheal vaginations (*TrInv*), showing the formation of blood cells (*BlC*), x 387.

changes in form.   The cells of the mesodermal tubes have lost
their columnar shape and from the inner layer a ridge now projects abutting against the yolk, directed mesiad and ventrad, the
splanchnic layer or rudiment of the enteric muscles (*MclEnt*).
Near the mid-line the yolk has receded a trifle and the mesoderm
as such has disappeared, leaving a narrow space between the
ectoderm and yolk, the *epineural sinus* (Heymons 1895) which is
partially filled with large loose cells, the blood corpuscles (*BlC*).

At Stage X (Fig. 79) the rudiments of all the mesodermal

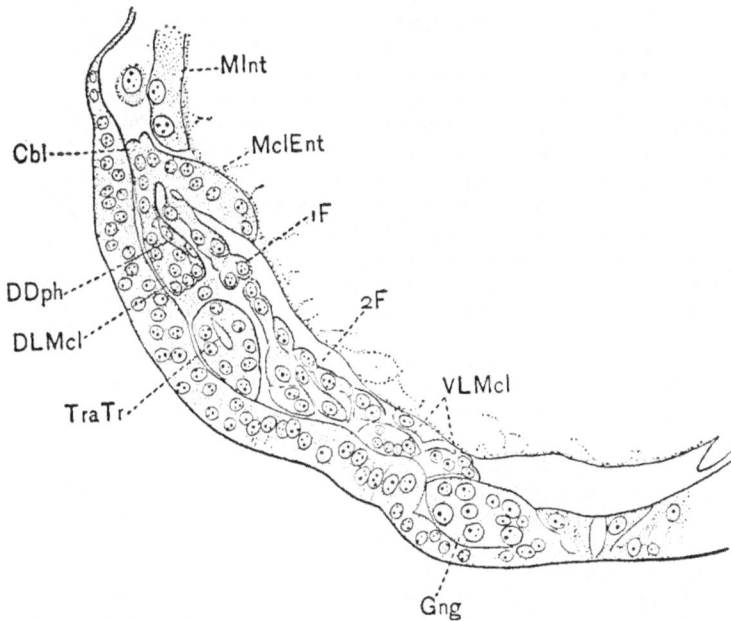

FIG. 79. Transverse section through the third trunk segment of an
embryo, Stage X, x 387.

organs and tissues may readily be identified.   The splanchnopleure
or rudiment of the enteric muscles (*MclEnt*) has changed from a
mere ridge to a well defined layer pressed close against the yolk.
The cells of the remainder of the visceral layer have for the most
part lost thir epithelial character and show a looser arrangement,
in anticipation of their destiny, which is to form the two main
divisions of the fat body (*1F, 2F*).   Moreover a break in the
visceral layer is seen opposite the rudiment of the tracheal trunk
(*TraTr*).   At the mesial edge of the mesoderm, close to the
ganglion of this segment (*Gng*), two groups of extremely small
cells are seen; these are the rudiments of the ventral longitudinal

muscles (*VLMcl*). The somatic layer of the mesodermal tubes now appears wedge-shaped in section, its broader end directed mesiad and ventrad, and is moreover split longitudinally into an inner thin single layer of cells, the rudiment of the dorsal diaphragm (*DDph*), and an outer portion composed of several layers of cells, which are to form dorsal longitudinal muscles (*DLMcl*). The yolk has continued to contract, drawing away from the mesoderm over its entire extent so that the epineural sinus is extended by the addition of two lateral cavities, which in turn are in connection with the coelomic cavity by means of the break in its visceral wall. These spaces together represent the rudiment of the definitive body cavity.

The process of differentiation goes steadily forward during the

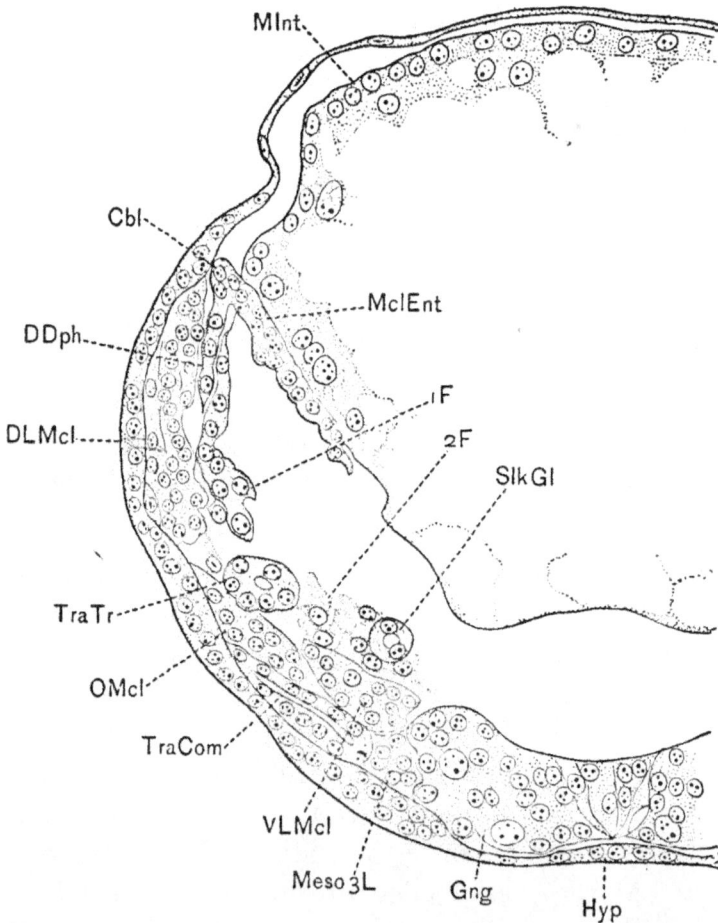

FIG. 80. Transverse section through the third trunk segment of an embryo, Stage XI, x 580.

next stage (XI). The rudiment of the enteric muscles (Fig. 80. *MclEnt*) increases in extent and decreases in thickness. The rudiments of the first and second sections of the fat body become more clearly defined, the first section almost losing its connection with the splanchnopleure (*MclEnt*), while the second breaks up into its constituent cells. The rudiment of the dorsal diaphragm now finally parts company with the remainder of the somatic layer of the mesodermal tubes, and together with the rudiment of the enteric muscles, forms in section a figure comparable to an inverted V. The apex of the V is formed by a group of small cells, the *cardioblasts* (*Cbl*), whose fate is that of forming the two halves of the heart. In the section figured the rudiments of three sets of muscles are plainly evident by reason of their pale appearance and small nuclei. The dorsalmost and largest of these is that of the dorsal longitudinal muscles (*DLMcl*), below this, lying between the tracheal trunk (*TraTr*) and the hypodermis, is the rudiment of the oblique muscles (*OMcl*), while next to this, ventrad of the silk gland (*SlkGl*) is the rudiment of the ventral longitudinal muscles (*VLMcl*). Between this last and the hypodermis, laterad of the ganglion of this segment is a small group of pale cells, which constitute the mesodermal portion of the leg rudiments (*Meso3L*). The yolk continues to contract and its ventral half assumes a three-sided form, and the future body cavity becomes greatly increased in extent. The rudiments of all the mesodermal tissues and organs are now well differentiated from one another and during this stage and the one following (XII), they increase in size and together with the hypodermis extend rapidly dorsad, this movement leading finally to the complete enclosure of the yolk by the germ band at Stage XIV. The most noticeable change at Stage XI is seen in the rudiment of the enteric muscles (*MclEnt*). At the preceding stage this rudiment covered only an insignificant part of the yolk. The rudiment of the mesenteron (*MInt*) has, since Stage IX (Fig. 78) covered the dorsal side of the yolk, but not until Stage XI does it come into contact with the rudiment of the enteric muscles. At this stage however it begins to extend rapidly ventrad over the yolk—which now is cylindrical in form—insinuating itself between the latter and the muscle rudiment. At Stage XII (Fig. 81,

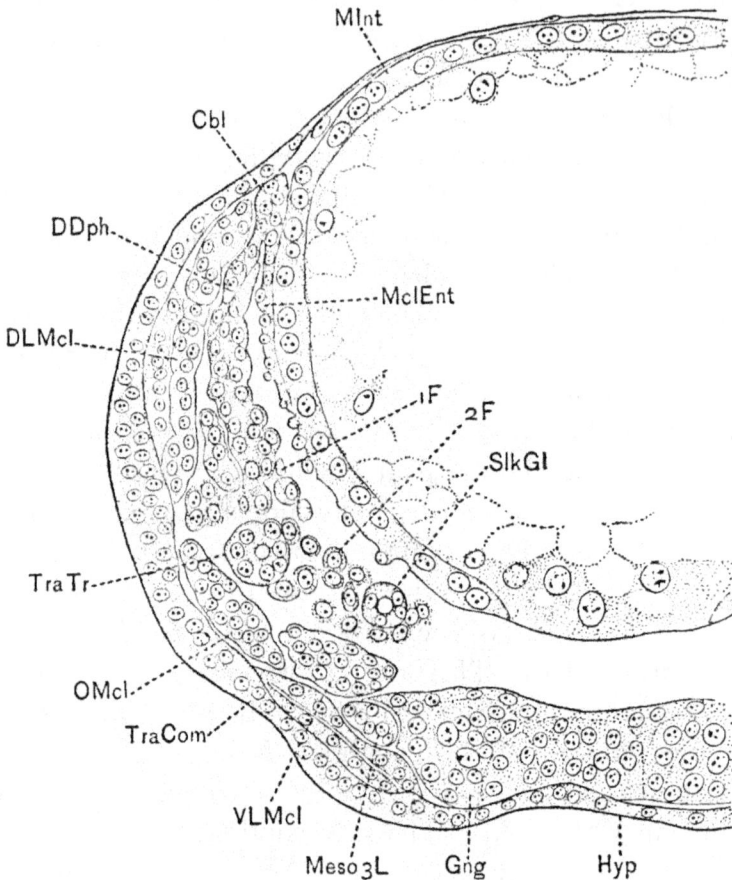

FIG. 81. Transverse section through the third trunk segment of an embryo, Stage XII, x 387.

MInt) it covers the lateral face of the yolk and has now reached the ventral surface. The rudiment of the enteric muscles has followed this ventrad growth, but only imperfectly so. A short distance back from the ventral edge of the rudiment of the mesenteron are four scattered muscle nuclei, while dorsad of this the forming muscle cells form a continuous layer which is joined to the cardioblasts. The appearance of the section suggests that the cells of the rudiment of the enteric muscles had been forcibly dragged apart by the rapidly growing epithelium of the mid-intestine.

A section through an embryo of Stage XIII-XIV (Fig. 82) shows the process of differentiation nearly completed. The hypo-dermis extends on each side up to the dorsal mid-line. The

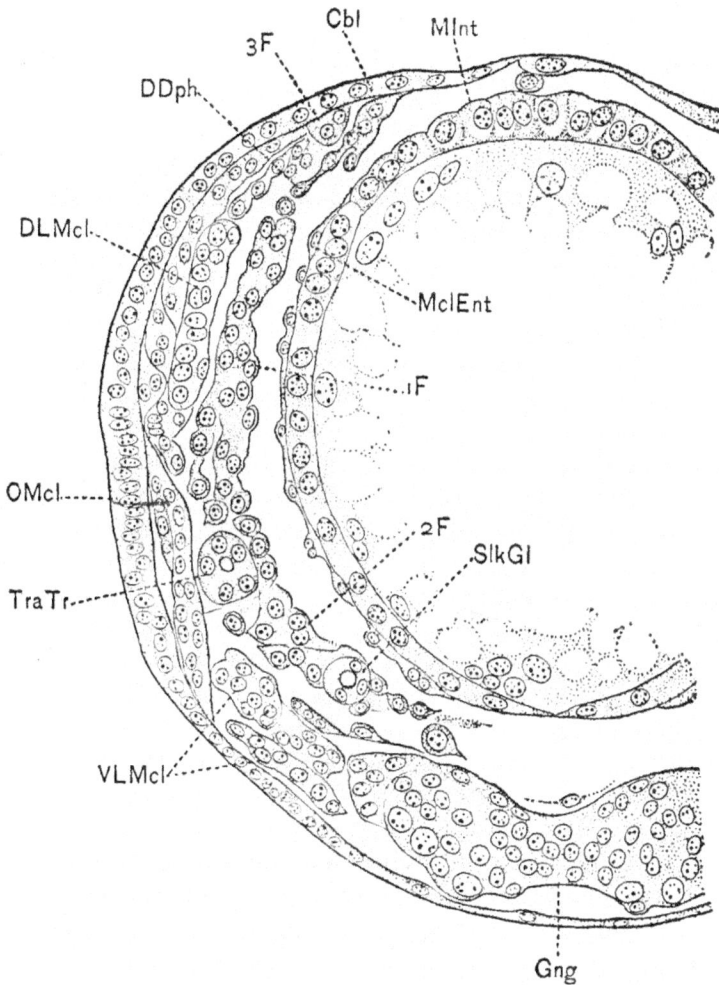

Fig. 82. Transverse section through the third trunk segment of an embryo, Stage XIII-XIV, x 387.

cardioblasts (*Cbl*) however are still some distance laterad of this point, but have lost their connection with the rudiment of the enteric muscles, which now uniformly cover the lateral surface of the mid-intestine (*MInt*). Above the dorsal diaphragm and laterad of the cardioblasts are two small cells evidently not muscle cells; these are fat cells belonging to the third section of the fat body (*3F*) or pericardial fat cells. These are derived from the dorsal margin of the somatic layer of the mesoderm. The other tissues and organs derived from the mesoderm have virtually attained their final form, and may be recognized without difficulty.

The final stages leading to the completion of the heart are shown in figure 83A and B, representing cross sections of the region

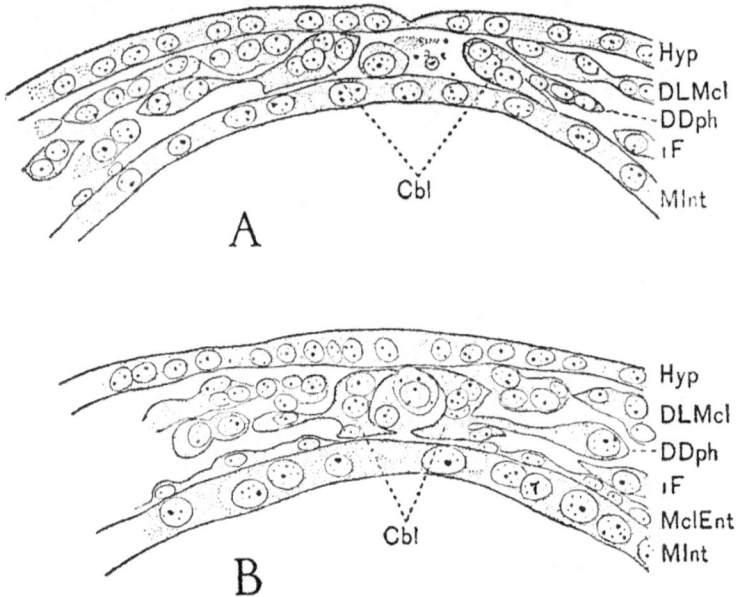

FIG. 83. Transverse sections through the dorsal region of the fourth trunk segment of late embryos, illustrating the final stages in the development of the heart. A, Stage XIII-XIV; B, XIV, x 600.

of the heart in the fourth trunk segment of Stages XIII-XIV and XIV. In the first figure (83A), the cardioblasts (*Cbl*) form two strips or strands of cells, each situated a short distance laterad of the mid-line, and with their mesial edges directed slightly dorsad. The rudiment of the dorsal diaphragm (*DDph*), consisting of a single layer of cells, projects on each side from the lateral border of the cardioblasts. On the right hand side of this figure, a narrow ridge projects out from the cardioblasts, just below the dorsal diaphragm. This is the last vestige of the former union of the cardioblasts and the remainder of the splanchnopleure. In the next figure (83B) the mesial edges of the cardioblasts of each side have approached one another until they are on the point of meeting. The former lateral margins of the cardioblasts have meanwhile also moved rapidly mesiad and have been turned about or "tucked under," so that they also are now directed mesiad and almost in contact with one another. In transverse sections therefore the cardioblast strand of each side is crescentic in outline. The formation of the heart is obviously completed by the fusion of the two cardioblast strands (see Fig. 76).

The two rudiments of the dorsal diaphragm (*DDph*) are seen to be formed from a single layer of more or less rounded cells, originally of uniform size. These, at Stages XIII and XIV (Figs. 83A and B, *DDph*) are seen to have lost this uniformity, cells of two sizes being now present, some of them having nuclei whose size approaches that of the fat cells, while others have nuclei only about half as large. The larger cells moreover have a rounded contour, the smaller on the contrary are thin and flat. These are to be considered as constituting, in a strict sense, the diaphragm, while the larger rounded cells are the paracardial cells. This is in harmony with the results of both Heymons (1895) and Carrière and Bürger (1897) and justifies the statement made on a previous page in regard to the homology of these cells.

The ventral diaphragm is formed from muscle fibres arising near the ventral longitudinal muscles, which extend out toward the mid-line to join those of the opposite side.

The structure and mode of development of the mesoderm just described obtains throughout the region of the trunk, extending from the middle of the second maxillary segment to the thirteenth trunk segment. As has already been said, it is less modified than in the head or in the extreme posterior abdominal region. In the anterior region of the head, at Stages VIII and IX, there is found a mass of somewhat loosely arranged mesoderm cells filling the head cavity, the more central of which are applied to the surface of the stomodaeal invagination (see Fig. 52 *Meso*). Caudad of this region the mesoderm cells divide into two lateral groups, and these in turn give place in the antennal region to a pair of thin-walled sacs, each composed of a single layer of flat cells. These are the coelomic (mesodermal) sacs of the antennal segment (Fig. 84, *AntMeso*). Each of these coelomic sacs sends off laterad a solid prolongation into the cavity of the antennal rudiment, filling the latter like a plug. The antennal coelomic sacs diminish in size caudad and each finally terminates in a flattened strand of mesoderm which traverses the mandibular and first maxillary segment and in the second maxillary segment becomes

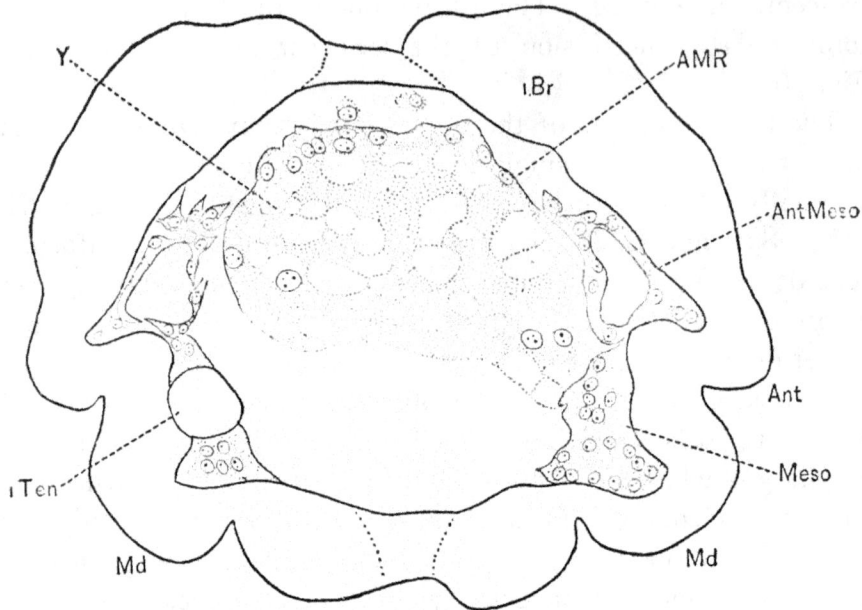

FIG. 84. Transverse section through the cephalic region of an embryo, Stage IX, intersecting the antennal rudiments (*Ant*) and their mesodermal sacs (*AntMesa*). Ectoderm shown in outline only, x 243.

continuous with the mesodermal tubes. The cavities of all of the rudiments of the gnathal appendages are stuffed full of mesodermal cells, which in the first and second maxillary segments are joined to the lateral strand on each side by a thin bridge of cells (Fig. 67, *Meso*).

The fate of the mesodermal elements of the head is briefly as follows: The central part of the anterior mass of mesoderm enveloping the stomadaeum becomes the layer of oesophageal muscles. The remainder of this mass is consumed in the production of those muscle strands which traverse the labrum and adjacent parts of the head capsule. The coelomic sacs of the antennal segment, at first quite small (Fig. 84, *AntMeso*) expand rapidly as the yolk is withdrawn from the head region and at Stage X become thin walled sacs which nearly fill the posterior half of the head capsule. Their mesial walls are however thicker than the lateral walls. As these sacs expand they acquire, in tranverse section, a more or less crescentic outline, their concave sides being turned towards one another and enclosing the future oesophagus (Fig. 58A and B, *AntMeso*). At Stage XIII the extreme mesial edges of these coelomic sacs, corresponding to the dorsal points of the crescents, meet and coalesce in the posterior region of the head, forming here an annular blood lacuna around the oesophagus (Fig. 42). On the dorsal side the line of coalescence extends back to and joins the dorsal wall of the heart, so that the latter empties into the annular lacuna, which now constitutes the aorta. This condition obtains only in the extreme posterior region of the head, since the ventral line of coalescence is short, so that the aorta soon comes to open on its ventral side into the general cavity of the head (Fig. 58C). The history of the remainder of the cephalic mesoderm offers little of interest. The mesoderm cells increase by division and arrange themselves gradually into masses corresponding in form to the various muscles of the head, which they are destined to produce. The mesodermal mass which constitutes the principal adductor muscle of the mandible is seen in figure 52 *Meso*. With the exception therefore of the antennal coelomic sacs, the mesoderm of the entire head, including the gnathal region, is consumed in the production of muscular tissue.

The mesoderm of the posterior trunk segments, that is beyond the eleventh trunk segment, has not been studied in detail, owing to the difficulty presented by the flexure of the posterior end of the embryo about the caudal pole of the egg. A flattened mass of mesoderm cells was however observed lying close to the anterior side of the proctodaeum (with reference to the axis of the embryo), at the time of appearance of the latter. This mass probably belongs to the terminal or fifteenth trunk segment. At a slightly later stage the proctodaeum is enveloped by mesoderm cells, evidently derived from this mass, and this envelope of mesoderm obviously represents the rudiment of the muscular layer of the proctodaeum. This is in agreement with the observations of Heymons (1895) on *Forficula*.

The successive phases of development undergone by the mesoderm, differ to a considerable degree among the representatives of the different orders of insects, nevertheless a fundamental similarity exists, and in the development of the mesoderm of the majority of insects, subsequent to its establishment as a germ layer, a common type is discernible, more or less modified, according as the insect—speaking generally—is specialized or primitive. In the myriopods, which probably stand as close to the ancestors of the insects as any existing group of the arthropods, the mesoderm, shortly after its formation, becomes broken up into segments, the mesodermal somites, and these in turn give rise to a series of paired sacs, the *coelomic sacs,* a pair of these corresponding typically to each segment. Each of these sacs sends out a hollow prolongation into the appendage of the corresponding side of the segment. The presence of paired coelomic sacs is probably primitive since it is also characteristic of the annelid worms. A precisely similar condition is found in the trunk region (thorax and abdomen) of *Lepsima* (Heymons 1897) and several members of the Orthoptera (see Heymons 1895). In all these forms the walls of the mesodermal sacs are thin and the various derivatives of the mesoderm are not formed *in situ,* as in the honey bee and many other forms, but from infoldings of the walls of the sacs. In *Forficula* (Heymons 1895) the coelomic sacs have thicker walls than in the forms previously mentioned and their derivatives are formed directly from them, without the intervention of folds. In the Coleoptera (*e.g. Hydrophilus* Heider 1889) a still more modified condition exists. Here the coelomic sacs of the trunk are thick-walled, and their lumen is relatively small, being restricted to the extreme lateral region of the mesoderm, a condition recalling that found in the honey bee, except that in the latter the cavities are continuous longitudinally, constituting the mesodermal tubes. In the muscids (Graber 1889) the cavities of the mesodermal somites are entirely absent, the mesoderm being solid throughout. This is without question a highly specialized condition.

In the majority of instances the cavities of the coelom are formed secondarily by clefts which appear in the previously solid mesoderm. Heider (1889) however has maintained that the coelomic cavities are produced by a separation of the two layers of mesoderm previously formed by the longitudinal folding of the blastoderm (see p. 49). This view was strongly opposed by Graber (1891) and has since not been confirmed by other investigations on the development of the Coleoptera; on the other hand Carrière (1890) has found that the coelomic cavities in *Chalicodoma* are formed in precisely this manner.

In all cases there is a median layer of mesodermal cells connecting the two lateral rows of mesodermal somites.

In the honey bee, the mesoderm, as described by Kowalevski (1871), Grassi (1884) and the writer, the coelomic cavity is represented by a narrow cleft, near the lateral margins of the mesoderm. Here the splanchnic and visceral layers are well defined and thick, being composed of long epithelial cells, but mesiad of this point these layers become rapidly thinner and are poorly defined. In *Chalicodoma* (Carrière and Bürger 1897) this distinction is even more sharply marked (see p. 197). In the honey bee no coelomic sacs, as such, are distinguishable, the two mesodermal layers on each side being continuous, longitudinally, as well as the cavity bounded by them. In *Chalicodoma* the thin-walled portion of the mesoderm, lying mesiad of the mesodermal tubes is distinctly divided into segmentally arranged sacs ("meso-dermal sacs"), while Bürger (1897) found that the mesodermal tubes were at a certain stage divided by faint partitions into chambers corresponding to the "mesodermal sacs." In the case of this insect therefore well developed coelomic sacs are present, although flattened dorso-ventrally so that the somatic and visceral layers are in contact with one another, except at their lateral margins. Here the cavities of the coelomic sacs are best developed but are on the other hand virtually fused in a longitudinal direction. In the honey bee this fusion has progressed to such an extent that the adjoining walls of the coelomic sacs of each side have been completely lost and their dorsal and ventral walls have become continuous throughout the entire length of the trunk.

As regards the fate of the different parts of the trunk meso-derm, modern investigators are in fairly substantial agreement. From the outer or somatic layer of the coelomic sacs are produced the trunk muscles and the major portion of the fat body, from the inner or visceral layer are produced the muscles of the mid-intestine and the genital ridges. The heart is formed from cells situated at the external margin of the coelomic sacs, where an angle is formed by the junction of the somatic and visceral layers. The median layer of mesoderm forms the blood cells.[21]

---

[21]Nusbaum (1886, 1888), Nusbaum and Fulinski (1906) and Hirschler (1905, 1909, 1909a) claim that a portion of the mid-intestine is formed from this median strip, which is therefore regarded by these investigators as entoderm (see pp. 73).

The formation of the definitive or secondary body cavity is inaugurated by the appearance of the epineural sinus, which is produced mainly by a withdrawal of the yolk from the embryo along the ventral mid-line. As the yolk continues to withdraw from the embryo the epineural sinus becomes extended laterally. Next the dorsal or visceral wall of the coelomic sacs becomes broken through so that their cavities become continuous with the epineural sinus, thus forming the definitive body cavity.

The development of the mesoderm in the honey bee conforms to the above ideal scheme quite well, except in one rather important particular: the fat body—with the exception of the pericardial fat cells—is formed, not from the somatic wall of the coelomic sacs, but from their visceral wall, or rather that part of it not used up in the production of the enteric muscles. This peculiarity is also shared by the mason bee, *Chalicodoma*, so that it seems probable that it may be peculiar to the Hymenoptera in general. There appears to be no evident reason for this divergence and it would be futile to speculate concerning it.

In that region of the insect embryo which is to constitute the definitive head, the segments in this region are naturally much more modified than in the trunk. In *Forficula* (Heymons 1895) besides a paired mass of mesoderm in front of the stomodaeum, there is a well developed pair of coelomic sacs in the antennal and the three gnathal segments, and in addition a pair of unmistakable, although reduced coelomic sacs in the premandibular segment. In other Orthoptera (see Heymons 1895) a similar condition obtains, except that the coelomic sacs of the premandibular segment are reduced to mere groups of cells. In the embryos of the Coleoptera the accounts of the mesoderm of the head are somewhat conflicting. Heider (1889) states that in *Hydrophilus* the coelomic sacs are wanting in the cephalic region, but appear suppressed in the mandibular segment, and their development is delayed in the first maxillary segment. In the most recent account of the development of a Coleopterous insect (*Donacia*), Hirschler (1909) finds, in the cephalic region, coelomic sacs only in the intercalary (premandibular) and second maxillary segments. In the Hymenoptera, according to Carrière and Bürger's (1897) description of the head mesoderm of *Chalicodoma* (pp. 392-393):

"In an embryo, in which the mesoderm tubes are present we can generally follow the mesoderm forward into the antennal segment. We can determine, that it fills the antennae and all the jaw rudiments, inclusive of the premandibular rudiment, forming their cores. Its greatest development is attained by the mesoderm in the sides of the head segments, where it is represented by several cell layers, it is wanting in the middle or only forms, as in the antennal and premandibular segments, a small bridge consisting of a single sheet of scattered cells.

Possibly even before the appearance of the cavities which represent the beginnings of the primitive body cavities, a cleft becomes evident in the head region of the germ band, between ectoderm and yolk (precisely between the rudiments of the ventral ganglionic chain and the yolk), in this cleft the mesoderm cells are much scattered. This space later is continuous with the epineural sinus, and the mesoderm cells scattered, that is floating in it, are to be regarded as blood cells. It later extends greatly, although the mesoderm contained within it can scarcely be said to increase and in general remains limited to the appendages and their bases. Therefore in the *head section of Chalicodoma there is no formation of coelom*. The antennal segment however is an exception.

A little later coelomic cavities appear in the antennal segment. This condition is virtually paralleled in the honey bee.

The fate of the head mesoderm as gleaned from the various accounts, is briefly as follows: In *Forficula* (Heymons 1895) the mesodermal mass anterior to the antennal segment forms the muscular coat of the fore-intestine. The antennal coelomic sacs become greatly extended, particularly in a longitudinal direction, while their mesial walls become thickened, and finally join in such a way as to form a tube, the aorta, continuous with the heart at its caudal end. The mesoderm in the remaining segments is consumed in the production of muscles, except in the premandibular segment. Here the mesoderm forms the "suboesophageal body" described by Wheeler (1893). In the Coleoptera as represented by *Donacia* (Hirschler 1909) the muscular coat of the fore-intestine is apparently formed by the mesoderm situated in the anterior head region, judging by the figures. The aorta is formed by coelomic sacs in much the same way as in *Forficula*, except that Hirschler confidently asserts that these belong not to the antennal, but to the premandibular segment. In *Chalicodoma* the course of events is much like that in *Forficula*, except that no suboesophageal body is formed in the premandibular segment. To this the behavior of the head mesoderm of the honey bee corresponds closely. The only difference of any consequence is in the formation of the aorta. In *Chalicodoma,* as in *Forficula,* late in the development the walls of each coelomic sac become apposed, obliterating the lumen. It is not clear however that a tubular lumen is formed, but from the figures given it appears that this is not the case, but that the aorta is simply a lacuna or blood space bounded above by the walls of the antennal coelomic sacs, but opening below into the cavity of the head.

XII

SEX ORGANS—THE OVARIES

The rudiments of the ovaries, in the newly hatched larva consist of two elongate masses of cells, situated close to the dorsal surface of the mid-intestine, near the mid-line, and extend through the seventh to the ninth trunk segments inclusive. In their general form the ovarian rudiments are elongate fusiform, compressed in a direction at right angles to the surface of the larva (Figs. XV, 85, *Ov*), and are about twice or three times as wide as thick. The contour of the rudiments is however actually irregular, as the figure shows. In transverse section the outline of the ovarian rudiments is, generally speaking, that of an elongate ellipse (Fig. 85, *Ov*). Six to ten rows of nuclei are included in the long axis of the ellipse, in its short axis not more than two rows, and frequently only one. A somewhat irregular layer of flattened cells surrounds the ovarian rudiment, and constitutes the rudiment of

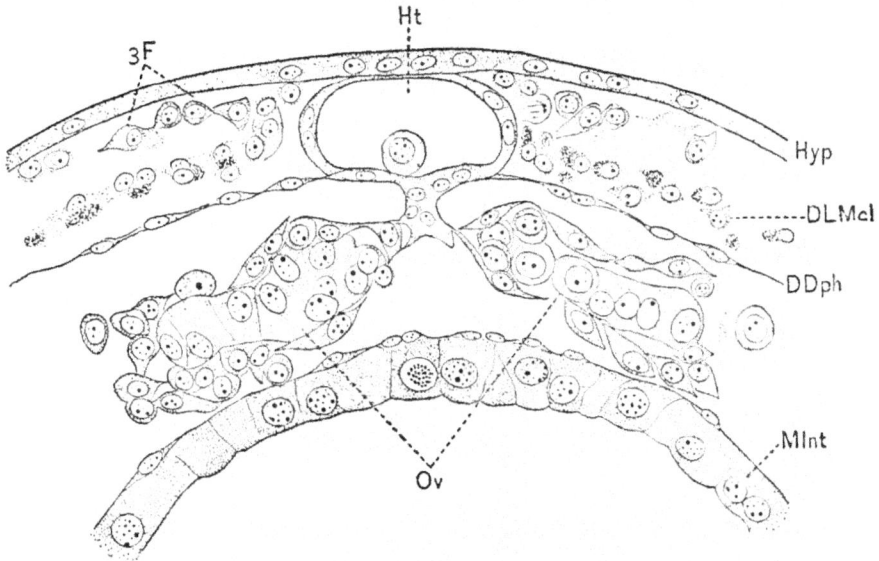

FIG. 85. Dorsal part of a transverse section through the posterior half of the eighth trunk segment of a newly hatched larva (Stage XV), intersecting the ovaries (*Ov*), x 567.

213

the ephithelial envelope of the ovary. The cells of this layer are joined together to form an epithelium, but as yet only an imperfect one, since there are frequent gaps between the cells, and there are also many cells destined to form epithelium, which have not yet become completely differentiated. This is evident in figure 85. At the mesial borders of the ovarian rudiments the ovarian epithelium is fairly well formed, and here it is united by a slender point of attachment to the ridge, previously described, which extends along the ventral side of the heart, thus holding the ovaries in place. The component cells of the ovarian rudiments present nothing peculiar in character. Their nuclei are large as compared with those of the adjoining tissues, approaching the size of those of the mid-intestine. As will appear later the cells of the ovarian rudiments are not perceptibly different from the undifferentiated mesoderm cells of earlier stages.

The ovaries of the honey bee are derived from the *genital ridge,* which is formed from the visceral wall of the mesodermal tubes in the fifth to the tenth trunk segments inclusive. At Stage X (Fig. 86A) the visceral wall of the mesodermal tubes has already become divided into two layers; an inner single layer of cells, lying in contact with the yolk, the rudiment of the enteric muscles, or splanchnopleure, and an outer, thicker portion, in which the cells are somewhat irregularly but compactly arranged and which constitutes a part of the visceral layer of the mesoderm. This is the portion destined to form the genital ridge (Fig. 86A, *Ov'*). In tranverse section it approximates an ellipse in outline, and in its narrower diameter includes from one to two layers of cells which in appearance do not perceptibly differ from those of the adjacent derivatives of the mesoderm. Even at this early stage the genital ridge has become narrowed at its point of attachment to the neighboring structures, the cardioblasts and the splanchnopleure.

Prior to Stage X, the cells destined to constitute the genital ridge may be seen in active mitotic division and during this period it increases slightly in thickness. Its nuceli seem also to increase slightly in size. At the next stage (XI) the remainder of the visceral layer, now composed of loosely arranged cells, becomes detached from the genital ridge. The outer or somatic wall of the mesoderm is meanwhile becoming differentiated into the rudi-

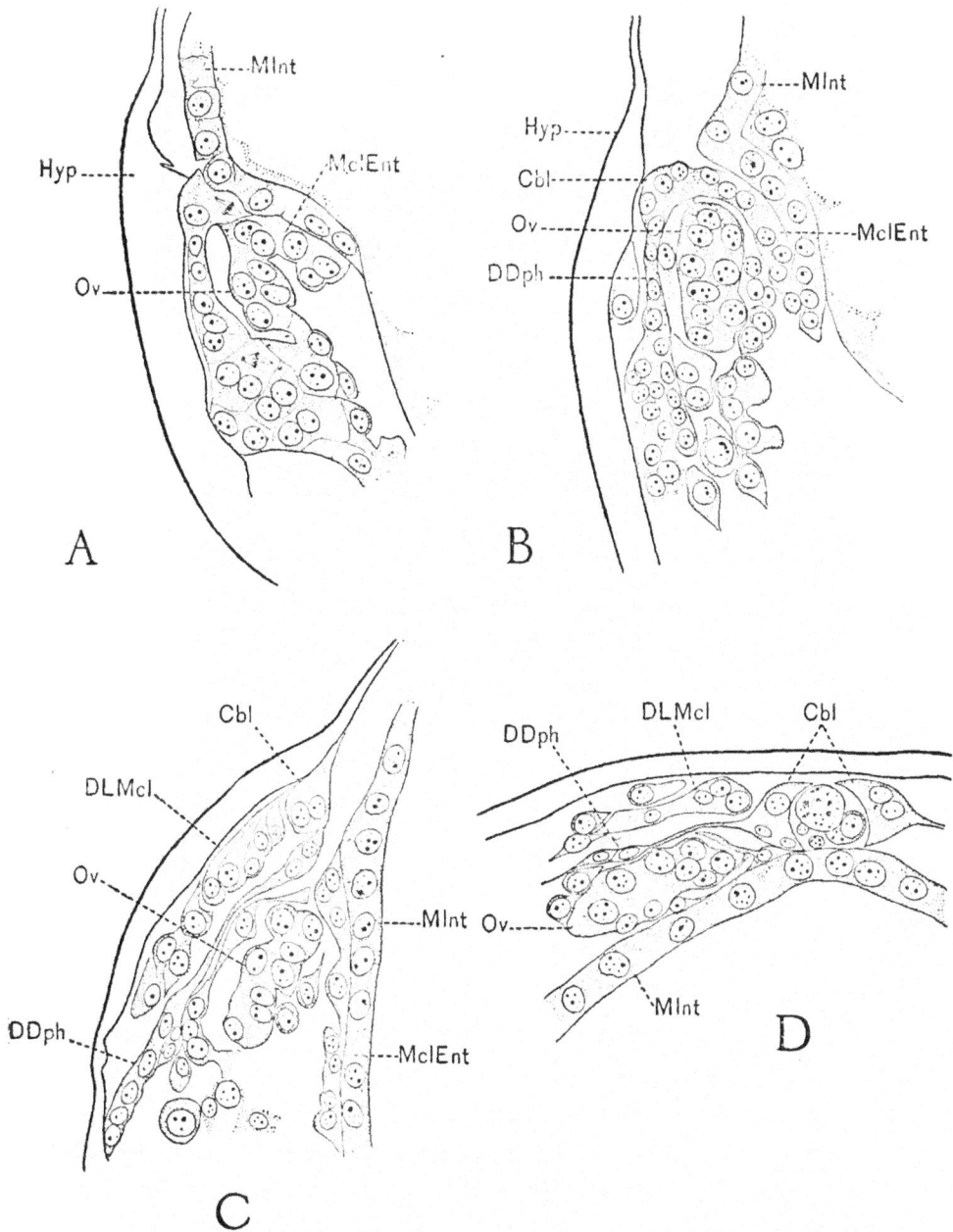

FIG. 86. Transverse sections through the left rudiment of the ovary (*Ov*) of four different stages. A, Stage X-XI; B, Stage XI; C, Stage XIII-XIV; D, Stage XIV, x 567.

ments of the dorsal diagraphm (*DDph*) and of the trunk muscles (*DLMcl*) while at the juncture of the two layers the cardioblasts (*Cbl*) are found.

The remaining cells of the visceral wall of the mesoderm mesiad of the genital ridge constitute a loose mesenchymatous tissue the greater part of which is transformed into fat cells. At an early period—Stage XI (Fig. 86B)—a few of the dorsalmost cells of this mass group themselves about the genital ridge, pushing up on both sides of it, and become applied to its outer surface preparatory to forming the epithelial envelope of the ovary (Fig. 86C and D).

At Stage X, the genital ridge, as has been said, extends from trunk segments five to ten inclusive. At about Stage XII it loses its connection with the cardioblasts and the splanchnopleure and during the succeeding stages it becomes gradually shorter and thicker (compare Fig. 86B, C and D), until at Stage XIV it extends only from near the anterior end of the seventh trunk segment into the anterior end of the ninth, which is its position at hatching (Fig. XV). Meanwhile the mesoderm cells grouped about the ovarian rudiments have gradually assumed the flattened form characteristic of epithelial cells and are plainly seen to be forming an envelope for the ovaries (Fig. 86, C-D). The attachment of this envelope to the heart takes place during Stage XIV.

The development of the genital organs of the honey bee was first studied by Bütschli (1870), but owing to imperfect technique, his observations are of little value, since he succeeded only in seeing and identifying the ovaries in a late embryo—about Stage XIII. The account of Grassi (1884), although brief, is more satisfactory. This investigator describes and figures the development of the ovaries, with the aid of sections, and finds that these organs are derived from the mesoderm. He seems however to have failed to observe their origin from the inner (visceral) wall of the mesodermal tubes. Moreover his statement that the genital organs extend almost from the fourth to the eighth abdominal segments is an evident error since at the stage in question (X or XI) they extend much further cephalad. Moreover, the epithelial covering of the genital organs was missed entirely. Petrunkewitsch has more recently (1901, 1902) devoted two papers to an account of the development of the sex organs of the honey bee. In these papers he describes conditions which are to say the least, somewhat peculiar and much at variance with the accounts given

by other investigators of the genital organs of insects. According
to this investigator the development of the male and female organs
of sex differ greatly. In the male the sex cells are derived originally from a cell produced by the fusion of the central product of
the division of the first polar body with the second polar body.
The cell thus produced gives rise by division to a considerable
number of cells which migrate in two groups to the dorsal side of
the embryo, where they again unite. From here they migrate
caudad, still increasing in number, to the abdominal region, where
they penetrate into the mesodermal tubes. Having arrived at this
point they become massed together to constitute the testes. The
ovaries, on the other hand, are formed from mesodermal cells, but
not from those of the mesodermal tubes, but from the loosely arranged cells derived from the mesoderm lying mesiad of the tubes.
An examination of Petrunkewitsch's figures (1903, Fig. 17) shows
that the mesoderm cells described as forming the ovaries are
identical with those which the present writer finds constituting the
epithelial envelope of the ovaries. The only remaining account of
the development of the genital organs of a hymenopterous insect,
aside from the more or less fragmentary accounts relating to the
parasitic forms, is in Carrière and Bürger's description of the
development of the mason bee (1897). In this form the rudiments of the genital organs first become evident as thickenings
of the inner or visceral wall of the mesodermal "sacs" of the third,
fourth and fifth abdominal segments. Since the sex rudiments
thus belong to the wall of the mesodermal "sacs," they are
situated mesiad (or ventrad) of the mesodermal tubes. The rupture of the visceral wall of the mesoderm takes place, as in the
honey bee, at the juncture of the mesodermal tubes with the remainder of the inner wall, consequently the sex rudiments are
separated from the mesodermal tubes and subsequently come to
lie free within the secondary or definitive body cavity. The
epithelial envelope of the sex glands is formed by mesodermal
cells which lie laterad of the rudiments of these glands, and it
is by means of this envelope that they become attached to the
heart, as in the honey bee. Toward the close of development the
rudiments of the sex glands contract longitudinally so that at the
time of hatching they lie entirely within the limits of one segment,

the fifth abdominal. Two points of difference are thus to be noted between the mason bee and the honey bee. In the former the rudiments of the sex glands are formed from the visceral wall of the mesoderm mesiad (ventrad) of the visceral wall of the mesodermal tubes, instead of from the visceral wall of the tubes themselves. Moreover the rudiment originally extends through but three of the abdominal segments instead of five, and contracts so as to lie in but one segment, instead of three.

Up to the year 1895 the belief prevailed among embryologists that the genital cells in insects were as a rule derived directly from mesoderm cells. To this rule however several important exceptions had already been noted; in all these cases the genital cells were derived not from the mesoderm, but from special cells set aside at an early period in the development. As early as 1865 Leuckart and Metschinkoff discovered that the genital organs in *Cecidomya* were formed from a group of cells (the "pole cells" of Robin and of Weismann) derived from a single cleavage cell originally situated at the posterior pole of the egg and readily distinguishable from the other cells during the early stages of cleavage. This discovery was confirmed by Balbiani (1882, 1885) and Ritter (1890) in *Chiromomus*. Metschnikoff as early as 1866 found that the cells destined to form the genital organs were distinguishable at an early period of development in *Aphis*, and this observation has been confirmed by later investigators (Witlaczil 1884, Will 1888a), while Woodworth (1889) and Schwangart (1904) obtain similar results in the case of the butterflies. In 1895 Heymons described the sex cells of *Forficula* and several members of the Orthoptera as being readily distinguishable at about the time of the formation of the mesoderm, and located at the posterior end of the germ band. They subsequently migrate cephalad along the somatic layer of the mesoderm to their definitive position. Soon after the publication of Heymons' researches, Lecaillon (1897a) announced that in certain chrysomelid beetles the sex cells could be distinguished during the formation of the blastoderm as a group situated at the posterior pole of the egg and differing from the ordinary blastoderm cells in their larger size and darker stain. Their subsequent migration is similar to that of *Forficula* and the Orthoptera. These results have been subsequently confirmed by several investigators (Friederichs 1906, Hirschler 1909, Hegner 1908).[22]

In the light of all of these observations, which relate to members of five of the orders of the insects, it seems probable that in all insects the germ cells are segregated at an early period. This is also in harmony with modern investigations on certain other animal forms, as for example those of Boveri (1887) on the round worm *Ascaris*, and Häcker (1897) on the crustacean *Cyclops*. In all of those insects, however in which an early segregation of the germ cells has been directly observed, these cells are more or less readily distinguishable by means of differences in size, clearness, etc. This is quite evident from the figures given by Wheeler (1893) and Heymons (1895), and is especially well shown by an original figure by Henneguy (1904, Fig. 378). In many insects the sex cells closely resemble the other cells of the embryo. This was found by Heymons to be the case in *Periplaneta orientalis* and in *Gryllus*. In these instances however the early segregation of the germ cells can be safely inferred by a comparison with closely related forms in which this difficulty was not encountered. In the honey bee there is, as a rule, little difference between the cells of different tissues and organs at the earlier stages. As has already been stated, the cells which are to constitute the ovary are at first indistinguishable from those of the mesoderm. It does not however follow that they are of mesodermal origin, even although they seem to constitute a portion of the mesoderm, since it is not at all unlikely that the germ cells may be set aside at an early period in development, and afterwards migrate into the visceral wall of the mesodermal tubes, and that such a migration may take place unobserved, on account of the similarity of the sex cells and mesoderm cells. A solution of this problem is perhaps to be found in the discovery of some constant difference hitherto unobserved, between the sex and other cells. An approximate solution is probably easier. If a closely allied form should be found in which the germ cells can be recognized without difficulty, then the behavior of the germ cells in the honey bee could be inferred by analogy.

[22] A complete review of the literature on this subject may be found in a recent paper by Hegner (1914).

XIII

ALIMENTARY CANAL

In the larva the alimentary canal comprises a short and narrow oesphagus, a capacious mid-intestine, and a short and but slightly curved hind-intestine. The mouth opening in the young larva is a transverse slit located just behind the labrum in the area between the bases of the mouth parts (see Figs. XIV and XV). At the posterior margin of the mouth is a flattened papillate process formed by the cephalad prolongation of the hypodermis forming the junction between the posterior (ventral) wall of the oeso-phagus and the ventral hypodermis. Just beyond the mouth the oesophagus dilates slightly to form a somewhat ill-defined pharynx, which is furnished with dilator muscles. From here the oesopha-gus curves uniformly dorsad and caudad and joins the mid-intestine just caudad of the point where the head and trunk join. The mode of junction is illustrated in figure 45, which shows that the fore-intestine (*FInt*) is invaginated into the lumen of the mid-intestine to form a structure corresponding in form to the proventricular valve of the imaginal bee. A similar oesophageal valve is found in many other insects. The lumen of the oesophagus, in its anterior portion, is crescentric in section, owing to the presence of a dorsal longitudinal fold (Fig. 41). In its posterior half four folds are present (Fig. 42), so that here the lumen has in section the form of a cross.

The mid-intestine is relatively large and occupies the greater part of the body cavity, extending from the first to the eleventh trunk segment (Fig. XV, *MInt*). Its form is that of an elongate cylinder with rounded ends. At its anterior end it communicates with the oesophagus, as just described, but its posterior end is completely closed. Its walls are relatively thin, but the cells composing them are somewhat larger than the majority of the other tissue cells. They are cubical in form, with a rounded nucleus (Fig. 75, *MInt*). The cytoplasm of these cells is granular and dark staining, and each cell commonly contains one or more large vacuoles.

The hind-intestine is a short tube, of about the same diameter as the oesophagus, bent in a gentle sigmoid curve, and traverses the three terminal trunk segments (Fig. XV, *HInt*). Its blind anterior end rests against the posterior end of the mid-intestine; its posterior end opens to the exterior at the posterior end of the trunk. Its structure is much the same as that of the oesophagus, except that its lumen is circular and relatively smaller.

The Malpighian tubules, four in number, open into the anterior blind end of the hind intestine (Fig. XV, *Mal*). From here they extend cephalad in a winding course to about the sixth trunk segment. Their diameter is slight, somewhat less than that of the lateral tracheal trunks, and is uniform throughout their length (Fig. 75, *Mal*). Their lumen is relatively small and circular in section.

The mid-intestine or mesenteron is derived exclusively from the anterior and posterior mesenteron rudiments. At Stage VII (Figs. 26B and 29, *AMR*) the anterior mesenteron rudiment is a thick convex disk of cells resting on the ventral side of the yolk, at the cephalic end of the egg. This disk is covered by the ectoderm with the exception of a small area at its posterior margin, where the cells of the disk extend through a gap in the ectoderm to the external surface (Fig. 26B). During the interval between Stages VII and VIII the germ band rapidly increases in length, the effect of this being to cause the cephalic end of the germ band to travel around the cephalic pole of the egg, carrying with it the anterior mesenteron rudiment, which assumes a position on the cephalo-dorsal face of the yolk. In this position the morphological posterior margin of the rudiment, where its cells reach the external surface, lies approximately at the cephalic pole of the egg. This rudiment, as a whole, has at first the form of a thick cap, but by the time Stage VIII is reached, it has become elongated in a longitudinal direction. Its form may then be compared with that of a spoon or scoop (Fig. 87A). The part corresponding to the handle is situated at the cephalic pole of the egg, where the stomodaeum (*Stom*) has already put in its appearance, and here it penetrates the ectoderm to the exterior, forming the bottom of the stomodaeal depression, while the remainder, corresponding to the bowl, extends over the dorsal surface of the yolk toward the caudal pole of the egg. Meanwhile the posterior mesenteron rudiment has also become altered in form. At Stage VII (Fig. 27C, *PMR*) it is disk-shaped, but between Stages VII and VIII it rapidly elongates, sending out a thin tongue-like process over the dorsal region of the egg toward the cephalic pole (Fig. 87B). At the same time the ectoderm grows caudad (with respect to the embryo) around the caudad pole, covering over that part of the posterior mesenteron rudiment which lies closest to the caudal pole of the egg (compare Figs. 27C and 87B, *PMR*). With the development of the tongue-like extension of the posterior mesenteron rudiment the rudiment as a whole becomes thinner, especially in the mid-line. The relatively rapid changes in form undergone by both mesenteron rudiments during Stages VII and VIII, appear to be due principally to changes in the form and arrangement of their component cells, and not to an increase in their number or volume.

In the example of Stage VIII represented by figures 87A and B, the anterior mesenteron rudiment extends over the dorsal surface of the yolk about one-third of the length of the egg. The thin tongue-like process of the posterior rudiment on the other hand extends towards the cephalic pole nearly one half of the length of the egg. The opposite edges of the two rudiments are therefore not far apart. During the interval between Stages VIII and IX both mesenteron rudiments continue to spread over the dorsal surface of the yolk until their opposite margins meet and fuse in the dorsal mid-line. The point of meeting is usually nearer to the cephalic than to the caudal end of the egg, on account of the more rapid growth of the posterior rudiment along the median dorsal surface of the yolk. By Stage IX the cells of the united mesenteron rudiments have become distributed in a single layer over the dorsal surface of the yolk (Fig. 78, *MInt, 88A, AMR*). The cephalic end of the yolk is also covered by the epithelium of the anterior mesenteron rudiment, part of which forms the floor of the stomodaeal invagination, as shown in figure 88A, *AMR*. The caudal end of the yolk on the other hand is still uncovered (Fig. 88B, *PMR*). The thickness of the newly formed epithelium of the mesenteron is at first not uniform, being greatest

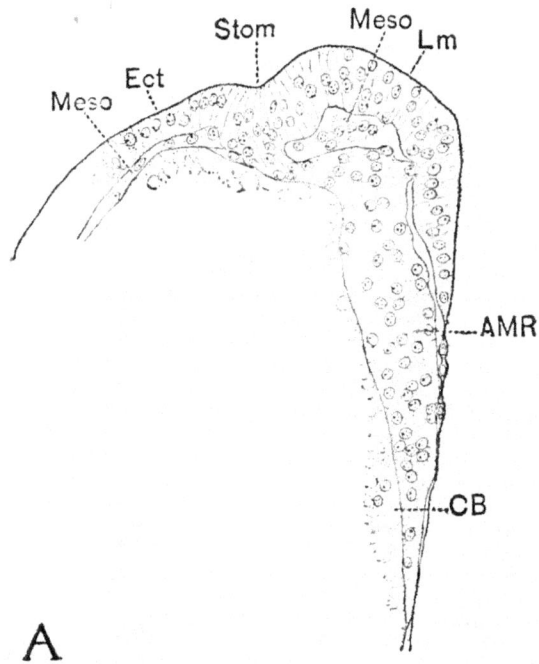

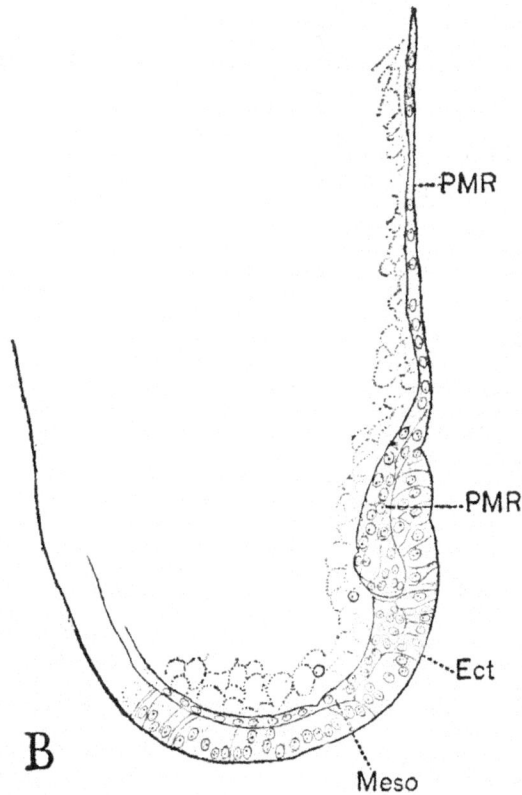

FIG. 87. Anterior (A) and posterior (B) ends of a median sagittal section through an embryo, Stage VIII, showing the anterior (*AMR*) and posterior (*PMR*) mesenteron rudiments, which are about to unite on the dorsal surface of the egg. The stomodaeal invagination (*Stom*) is just appearing. The relation of this to the anterior mesenteron rudiment is clearly shown, x 243.

at about the point of junction of the two rudiments. Near the posterior end of the yolk the epithelium is especially thin, as shown in figure 88B, *PMR*. At Stage X (Fig. 79 *MInt*) the posterior end of the yolk is covered by the future epithelium of the mid-intestine which now extends ventrad on each side to meet the splanchnic layer of the mesoderm. During the next three stages (Figs. 80, 81 and 82) the epithelium of the mid-intestine extends steadily ventrad on each side, until, at Stage XIII-XIV, the two edges meet and unite on the ventral mid-line (Fig. 82) thus completing the formation of mesenteron. The union appears to be virtually simultaneous throughout its extent. Meanwhile the rudiments of the fore- and hind-intestines, the stomodaeum and proctodaeum, are also developing. The stomodaeum puts in its appearance at Stage VIII in the form of a cup-shaped depression located at the cephalic pole of the egg. The floor of this invagination is formed by cells of the anterior mesenteron rudiment, as shown in figure 87A. Between Stages VIII and IX the depression deepens rapidly, becoming funnel-shaped, while at the same time the yolk retreats from the anterior end of the egg, its withdrawal keeping pace with the lengthening of the stomodaeum. The latter thus soon becomes a short tube, composed of a single layer of prismatic ectodermal cells, closed at its inner end by the cells of the anterior mesenteron rudiment (Figs. 52 and 60, *Stom*). During the earlier phases of the development of the stomodaeum, its cavity narrows gradually from its outer to its inner end, and the cells of the mesenteron closing the latter form a wall almost as thick as that of the stomodaeum itself, containing several nuclei (Fig. 60). Towards Stage XIV the inner end of the stomodaeum widens out, acquiring a flaring form, and the stomodaeal wall becomes much thinner just anterior to its junction with the mesenteron (89A). Careful examination of preparations of this stage shows that this condition is associated with the formation of a double fold of the stomodaeal wall preparatory to the invagination or intussusception of the inner end of the stomodaeum into the cavity of the mid-intestine. Correlated with these changes is a thinning of the cellular diaphragm-like wall closing the inner end of the stomodaeum, the central portion of this wall becoming reduced to a thin membrane, in which now no nuclei are

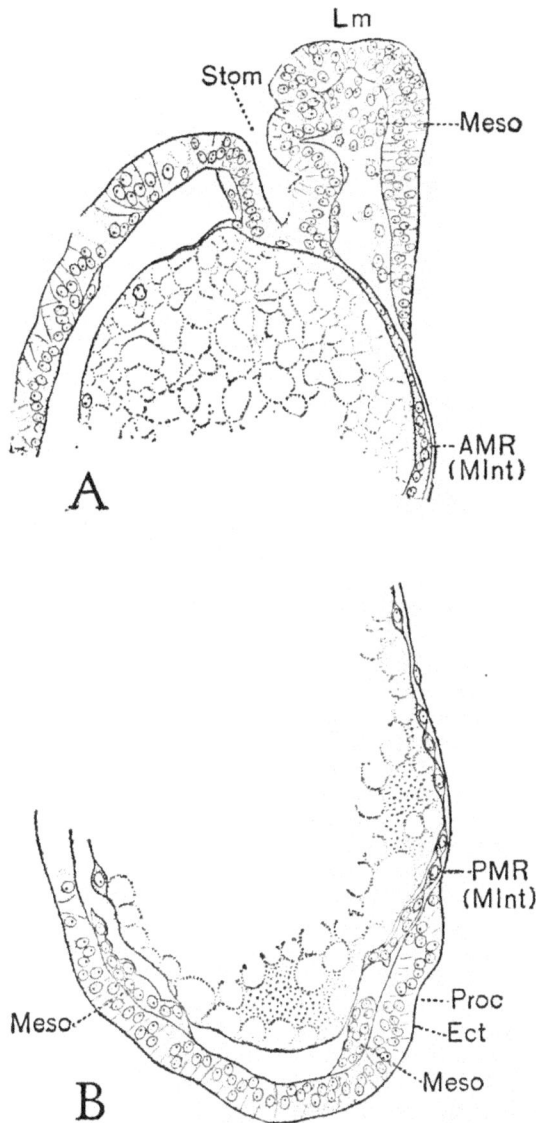

FIG. 88. Anterior (*A*) and posterior (*B*) ends of a median sagittal section through an embryo; Stage IX, showing the anterior (*AMR*) and posterior (*PMR*) mesenteron rudiments, which have now united along the dorsal surface of the egg. The stomodaeal invagination (*Stom*) is well developed, while the protodaeal invagination (*Proc*) is indicated by a slight depression, x 243.

present (Fig. 89A). The next and final step, which takes place about the time of hatching, involves the actual invagination of the inner end of the stomodaeum—now the oesophagus—into the mid-intestine, and the rupture of the closing membrane. This rupture seems to be caused mechanically by pressure of the inva-

ginating oesophageal wall; at least this is the impression given by
the sections studied, in which the peripheral part of the torn membrane is seen still attached to the wall of the mid-intestine(Fig.
89B). A similar proventricular valve appears to be formed in
much the same way in *Forficula* and *Periplaneta* (Heymons,
1895).

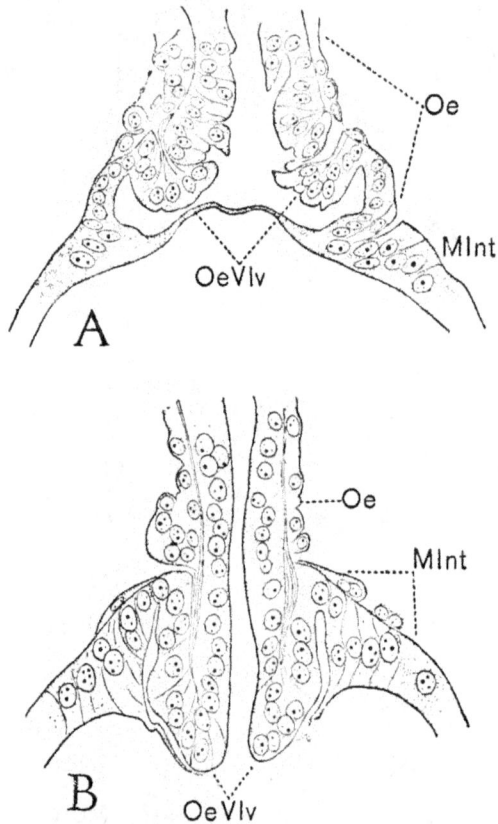

FIG. 89. Sagittal sections through the junction of the fore and mid-
intestines of two embryos, showing the formation of the oesophageal
valve (*OeVlv*). A, Stage XIV; B, Stage XV, x 567.

At the time of its first appearance the stomodaeum occupies a
position corresponding to the cephalic pole of the egg, and during
its earlier stages its lumen coincides with the long axis of the egg.
At Stage X the embryo begins to shorten, drawing the external
opening of the stomodaeum—the future mouth—to its final position on the ventral surface of the egg, the oesophagus therefore
acquiring a curved course. At the same time the hypodermis
forming the angle bounding the future mouth on its morphologi-

cal posterior side, lengthens out to constitute the papillate or
tongue-like process previously mentioned (p. 220). This is plainly
shown, although not lettered, in figure XIII, and lies behind the
labrum, bounded laterad by the mandibles and first maxillae.

The development of the different portions of the digestive
canal from their respective rudiments corresponds essentially
to that found in other insects having bipolar mesenteron rudiments, except in one particular. In all the other insect embryos
thus far studied, in which bipolar mesenteron rudiments occur,
including the mason bee, each of the rudiments sends out a pair
of epithelial bands, one on each side of the ventral mid-line. These
grow toward each other, meet and fuse. The narrow space left
vacant betwen them on the ventral side is next filled by cells derived from the bands. The ventral surface of the yolk is therefore
covered first, and the dorsal surface is covered subsequently by
gradual dorsad growth and extension of the bands. In the honey
bee, as Grassi (1884) discovered, it is the dorsal surface of the
yolk which is first covered by the extension of the mesenteron
rudiments; moreover, these extensions do not have the form of a
pair of bands. although certain statements made by Grassi suggest
this, so that Carrière (1897) has interpreted Grassi as stating that
each of the two mesenteron rudiments gives rise to "paired lateral
arms." Grassi's statement with reference to the anterior "entoderm" rudiment is as follows: "It is always more advanced
on the sides than in its median part, so that two or three
tranverse sections are always found in which it fails to occur
in the mid-line." Further on in the same chapter he makes a
similar statement regarding the posterior "entoderm" rudiment.
Examination of series of tranverse sections of Stage VIII have
failed to completely confirm these statements. During this period
of rapid growth the epithelial layers formed by the two rudiments
of the mesenteron are extremely irregular and variable in thickness. especially in the region of the rapidly advancing thin edges
on the dorsal side of the yolk. In two of the four preparations
of the same stage as figure 87 the statement of Grassi seemed
to hold, while in the other two the advance seemed greatest along
the mid-line. The point is however a trivial one since Grassi's
statements cannot be fairly construed to indicate the presence of
paired bands. Evidences of these, as such, seem to have almost

completely disappeared. The posterior mesenteron rudiment, gives perhaps the best indication of the former existence of paired bands in the incomplete bilateral subdivision of its thicker basal portion at Stage VIII .

The rudiments of the Malpighian tubules appear at Stage VIII, preceding, as Grassi (1884) discovered, the formation of the proctodaeum, which does not appear until Stage IX. The rudiments of the Malpighian tubules therefore for a short time open directly on the external surface of the embryo. A similar relation obtains in *Microgaster* (Kulagin 1892), *Leptinotarsa* (*Doryphora*) (*Wheeler* 1889), *Chalicodoma* (Carrière 1890, Carrière and Bürger 1897), and *Gasteroidea* (Hirschler 1909). It is only in *Apis* and *Chalicodoma* however that the Malpighian tubules are actually formed before the proctodaeum; in the others they appear on the external surface of the ectoderm around the already formed proctodaeum. In the honey bee the Malpighian tubules appear as four equidistant pit-like depressions at the extreme posterior end of the germ band, on the dorsal surface of the egg. According to Grassi's account the depressions are at first entirely distinct from one another, but subsequently the two lying on the same side of the mid-line merge with one another, a pair of longitudinal grooves being thus formed. Examination of preparations of embryos of the proper stage (VIII) failed to show the invaginations as distinct from one another; in all cases in which they were sufficiently well defined to be recognizable they had coalesced on each side of the median line to form a pair of slightly curved grooves, whose concavities faced one another (Fig. 90). Examination of sections shows also that the pair

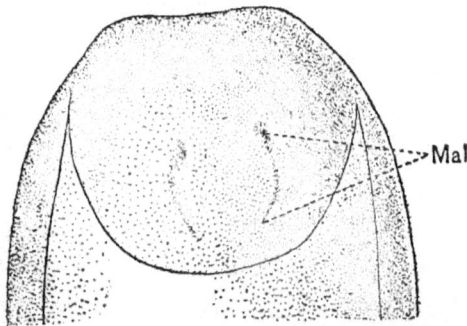

Fig. 90. Dorsal surface of posterior end of egg, Stage IX, showing the crescentic grooves from which the Malpighian tubules are formed.

of invaginations lying closest to the caudal pole of the egg—the future ventral pair of Malpighian tubules—are deeper than the anterior pair, and this difference is evident up to a late stage. At Stage X the ectodermal area included between the grooves described above becomes depressed, carrying with it the rudiments of the Malpighian tubules (Fig. 91, *Mal*). This area

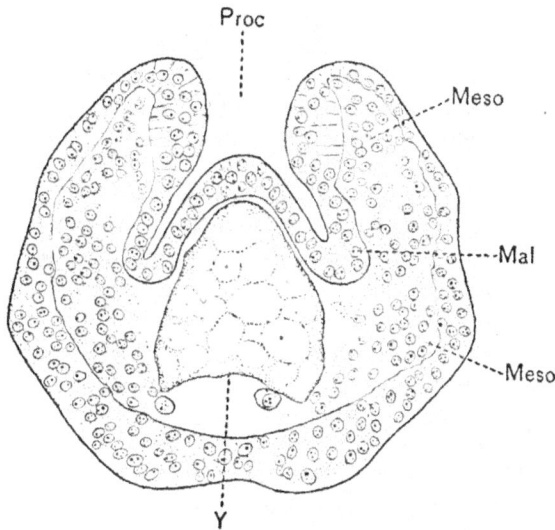

FIG. 91. Transverse section through the posterior end of an embryo, Stage X, intersecting the proctodaeal invagination (*Proc*). The posterior (ventral) pair of Malpighian tubules (*Mal*) are also shown.

therefore constitutes the floor of the proctodaeum. The opening of the latter is at first on the dorsal side of the egg, on account of the caudal flexure of the embryo, but with the gradual straightening of the embryo the external opening of the proctodaeum becomes terminal with respect to the egg. This takes place during the next stage (XI). From the time of its inception the proctodaeum elongates slowly and steadily, the yolk retreating from the posterior end at a corresponding rate. The inner end of the proctodaeum comes into contact with the epithelium of the mesenteron from the first, but the relation does not become more intimate than that of mere contact. The growth of the Malpighian tubules is extremely rapid; they push forward along the ventro-lateral wall of the mid-intestine as slender, nearly straight tubes, until their anterior ends reach the neighborhood of the sixth trunk segment. Here their cephalad extension seems in some way hindered, and their further elongation causes them to be thrown into a series of curves and loops (Fig. XV, *Mal*).

XIV

YOLK AND YOLK CELLS

In the chapter on "Cleavage" it is stated that not all of the cleavage cells migrate to the periphery of the egg, a few of these remaining behind in the yolk to form the yolk cells. In order to distinguish them from cells entering the yolk at a later period these yolk cells will be referred to as primary yolk cells, since they are primary in respect to the time of their origin. At first the primary yolk cells preserve the appearance of ordinary cleavage cells and divide simultaneously with their sister cells, which are about to form the blastoderm, exhibiting mitotic figures precisely like those of the cleavage cells (Fig. 93, A and B). Soon after the blastoderm has become established mitotic divisions of this character cease and are supplanted by mitoses of a more or less irregular character. In these the spindles are as a rule short and the chromosomes frequently agglomerated to form irregular masses (Fig. 93, C-G). This period lasts until after the nuclei forming the blastoderm have become arranged in two layers (18-20 hrs.). The large number of cells found in the yolk at this time is therefore to be attributed to mitotic divisions, since the latter are found abundantly until some time after the blastoderm has been formed. This includes both the earlier regular type as well as the later irregular type. Many of the mitotic figures of the later type resemble those found in pathological tissues—such as cancer cells—and often have the appearance of being unequal (Fig. 93D), or multipolar (Fig. 93, F and G), but the mitotic figures are so minute and the surrounding cytoplasm often so dark that a critical study of them was not attempted. The probability that both the unequal and multipolar types exist is strengthened by the subsequent frequent presence of nuclei of different sizes within the same mass of crytoplasm (Fig. 92A). These irregular figures might be considered actually pathological or abnormal, were it not for their great abundance in several different preparations of eggs of the

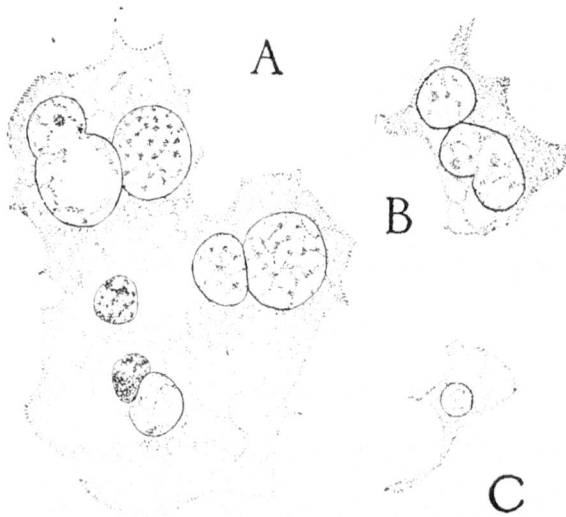

FIG. 92. Yolk cells from eggs eighteen to twenty hours old. A, large multinucleate cell, showing degenerating nuclei. B, small binucleate cell; one of the nuclei appears to be dividing by amitosis. C, small uninucleate cell, x 840.

same age, which are in all other respects perfectly normal. Moreover Lecaillon (1897a) has reported irregular mitotic figures in the chrysomelid beetle *Clytra* during cleavage. It is true that these figures, as stated in an earlier chapter may quite possibly be caused by too slow a fixation and are therefore to be considered actually abnormal. Friederichs (1906) however reports that irregular mitotic figures are of constant occurrence in the yolk cells of another chrysomelid beetle (*Donacia*) and are therefore strictly normal. This case is precisely parallel to that of the bee. Dickel (1904) also reports the occurrence of mitotic figures in the yolk cells of the honey bee, but neglects to say at what stage they are found. The single small figure given by him (his Fig. 4) indicates however that this is one of the small irregular spindles which succeed those of the regular type.

Examination of the text and figures of Petrunkewitsch's two papers (1901 and 1902) indicates that he also probably observed the small type of mitotic figures in the yolk cells but without recognizing their nature. The "double nuclei" in the "Richtungsplasma" have much the appearance of the small mitotic figures when seen at a relatively low magnification. On page 28 of the first paper (1901) occurs the following significant paragraph: "I should here also mention an observation. At the

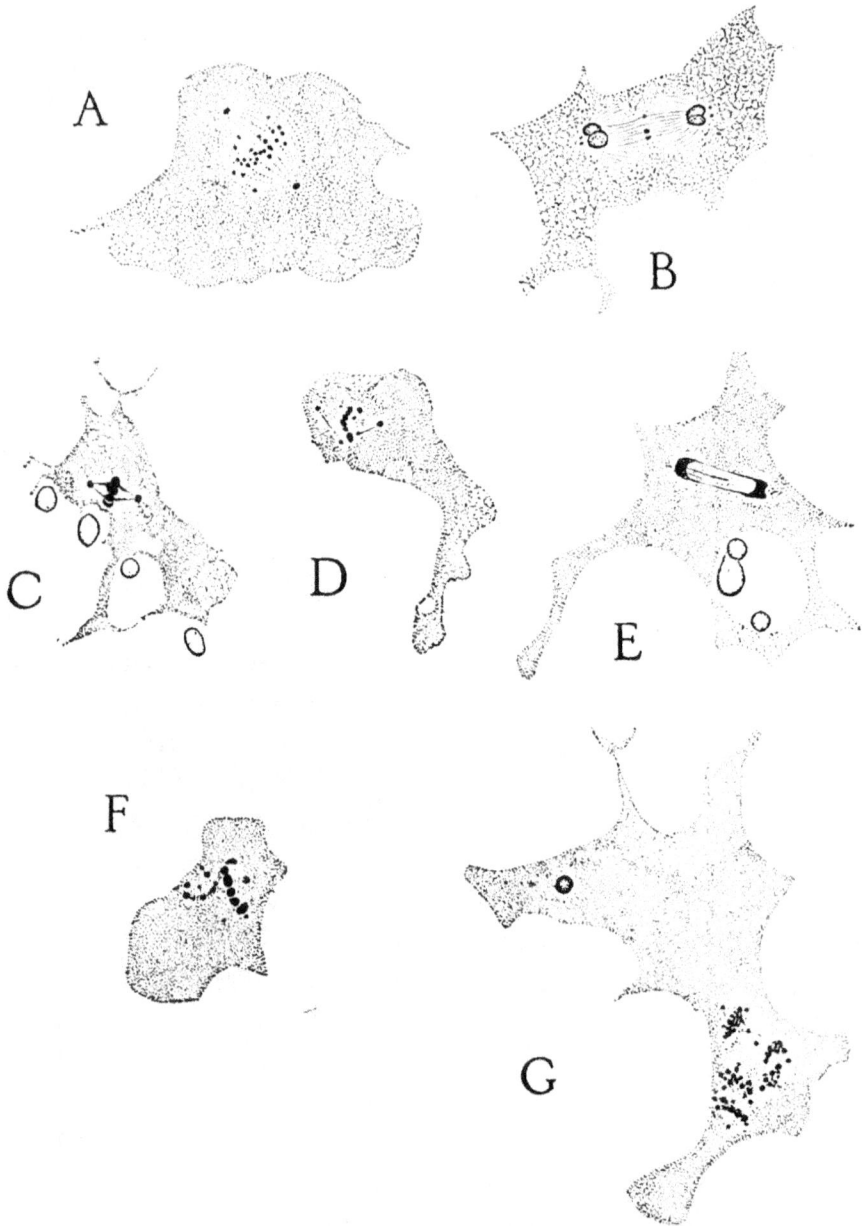

FIG. 93. Yolk cells, showing mitotic division. A, cell showing a regular symmetrical spindle of the ordinary (normal) type. B, telophase of the same type. C, cell showing small but symmetrical spindle. D, cell showing a small and slightly unsymmetrical spindle. E, cell showing chromatin aggregated into dense masses; telophase. F, G, irregular and asymmetrical spindles, x 1107.

time when the blastoderm has just been completed, the yolk is almost free of cleavage cells. These are found isolated only here and there. However when the head fold[23] is formed, a multitude of these cells appear in the yolk, which are agglomerated to the number of 8, 9 or even more in a single mass; some of these nuclei are now like the double nuclei described above, although not so clearly divided into two halves. At this time I could no longer discover the double nucleated cells. Could those (just described) have arisen from these?" It seems highly probable that the imperfectly double nuclei just described are the small mitotic figures; the resemblance of these which are found in the true yolk cells to those seen in the "Richtungsplasma" is especially interesting.

During the period of irregular mitotic divisions of the yolk cells small and densely stained bodies appearing to be composed of nuclear material (chromatin) are found within the cytoplasm of the yolk cells (Fig. 93G, upper left hand side). The precise origin of these bodies is unknown, but it seems that like the similar bodies seen later, they represent degenerating nuclei, and are also probably in some way causally related to the irregular mitotic divisions.

In preparations of eggs eighteen to twenty hours old, when the nuclei of the blastoderm are arranged in a double layer over the greater part of the blastoderm, the cellular contents of the yolk will be found to present a richly varied appearance, and one which differs not a little in different eggs; nevertheless the following description will apply to the majority of examples of this stage. Scattered throughout the yolk are to be seen numerous irregular islands of dark-staining cytoplasm of the most various shapes and sizes containing one to several nuclei. These cells—or syncitia, since they are nearly all multinucleate—are most numerous along the central line of the yolk, and are also rather more numerous in the anterior than in the posterior half of the egg, although syncitia of considerable size are frequently to be seen lying in contact with the inner surface of the blastoderm. The range in size of the syncitia may be gathered by comparing figures 92A and 92C. In the anterior half of the egg, always near the central line of the yolk, there are usually one or two syncitia of

[23] Reference to the figures shows that by this is meant only the dorsal strip of thin blastoderm.

relatively enormous size, like that shown in figure 92A. In a crude and general way, both the number and size of the nuclei vary with the size of the syncitia to which they belong. The largest syncitia accordingly are commonly found to contain a considerable number of relatively enormous nuclei; as shown in figure 92A. All of the yolk syncitia possess branching processes which are continuous with the threads of the delicate protoplasmic network remaining within the yolk subsequent to cleavage, and are thereby placed in direct connection with one another. In a broad sense therefore the entire protoplasmic contents of the yolk forms a single syncitium, directly comparable to that of the cleavage cells.

A large proportion of the nuclei of the yolk syncitia—in the narrow sense—show plain evidences of degeneration. A survey of the preparations covering this period shows that the changes are as follows: The nucleus decreases in size and the entire contents of the nucleus assumes a darker shade, while the chromatin at the same time becomes gradually agglomerated into two or three irregular dense masses, which finally become united into one as the nucleus continues to contract. The last observed stage in this process is a densely stained and relatively minute subspherical mass, as shown in figure 94, A-D. A similar agglomeration of the chromatin in the degenerating nuclei of the yolk cells has been observed by Wheeler (1889) in *Leptinotarsa* (*Doryphora*) and by Friederichs (1906) in *Donacia*. Since the nuclear membrane is intact at all stages of degeneration, it necessarily follows that the karyolymph is lost by diffusion, passing out into the surrounding cytoplasm which often takes on a lighter hue about such nuclei. The masses of chromatin formed during nuclear degeneration are located in various parts of the nucleus, in some cases at the center, in some around the periphery, in others at one side. It thus happens that in some instances a densely stained spherule is seen lying on the internal wall of a small vacuole (Fig. 94E), the latter representing the empty space left within the nuclear membrane. In the majority of cases the degenerating nucleus for some unknown reason lies at first close to a normal nucleus, so that during the late stages of degeneration the appearance suggests that the products of degenera-

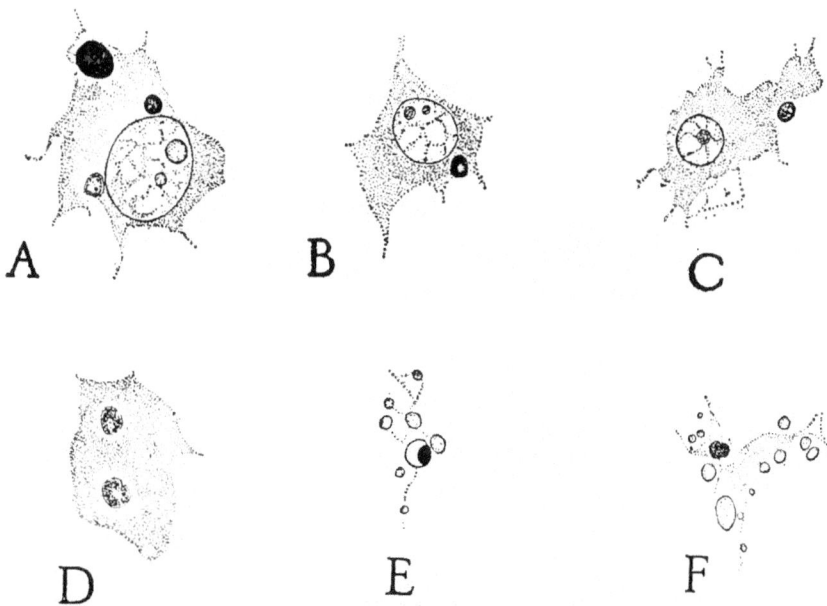

FIG. 94. Degenerating yolk nuclei. A, yolk cell, showing one large vesicular nucleus, and three nuclei in each of which the chromatin has become aggregated into a spherule. B, spherule leaving cell. C, spherule lying just outside of cell. D, two spherules lying inside body of cell. E, F, nuclei lying in the meshes of the protoplasmic reticulum, x 840.

tion, the spherular masses of chromatic material, may have emanated from the normal nucleus (Fig. 94, A and B). Friederichs (1906) in the course of his investigations on the embryology of the chrysomelid beetles, found that during the earlier stages of development the nuclei of the germ cells of the ectoderm and of the mesoderm, gave rise, by a peculiar kind of direct division to structures resembling degenerating nuclei, which he termed "paracytoids." These were not however thus formed in the yolk cells, but since the nuclei of the yolk cells form precisely similar bodies by ordinary nuclear degeneration, these too were termed "paracytoids." In spite of this last mentioned circumstance and in view of the frequent apposition of the dark-stained spherules and normal nuclei in the yolk cells of the honey bee, it seemed advisable to determine definitely whether any of these spherules were derived by the emission of chromatic material from adjoining nuclei. A careful examination under high powers of all the available sections of the stages during the formation of the blastoderm shows conclusively that there is no evidence of a direct genetic connection between the dark-stained spherules and the adjoining nuclei, since in every case a small but appreciable space separates the two kinds of nuclei. Moreover the nuclear membrane of the normal nucleus is always intact at every point. It may be assumed therefore that all of the sub-spherical deeply staining masses observed within the yolk are formed by ordinary degeneration of yolk cells. At the stages beyond eighteen to twenty hours up to the time of the formation of the germ layers, densely stained chromatic spherules are commonly found isolated within the yolk, sometimes appearing as though caught in the meshes of the protoplasmic network (Fig. 94E), sometimes surrounded by a fragment of cytoplasm (Fig. 94F). That these are derived from yolk cells is evident by comparing them with those still enclosed within the yolk cells (Fig. 94D). In addition, it is possible to find a spherule in the act of leaving a yolk cell (Fig. 94B), as well as other stages in the emigration of the spherules out into the yolk (Fig. 94, A and C). The chromatin spherules are therefore equivalent to the "entodermal paracytoids" described by Friederichs (1906) in *Donacia,* since they have precisely the same history.

Shortly after mitotic division in the yolk cells ceases, and when the nuclei of the ventral region of the blastoderm have become arranged in a double layer, (18-20 hours) not a few of these nuclei may frequently be detected in the central margin of the blastoderm, at some distance from their fellows. Others may be seen half in the blastoderm, and half in the remains of the peripheral protoplasmic layer lining the blastoderm while still others are entirely within this layer (Fig. 95). Favorable preparations show a considerable number of nuclei in all these stages. The migrating nuclei apparently carry with them little or no cytoplasm and after leaving the blastoderm become embedded in the dark-stained vestiges of the peripheral layer of cytoplasm. Henceforward they can no longer be distinguished with certainty from the smaller representatives of the primary yolk cells. The yolk cells thus produced by migration from the blastoderm may be termed the *secondary yolk cells*[24] by reason of the later time of

[24] This term has already been applied by Cholodkowsky (1891) to certain cells in the cockroach embryo which he believed were derived from the

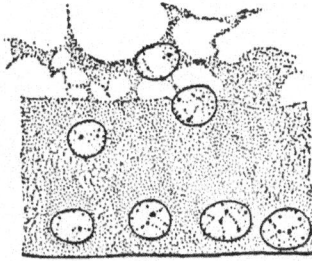

FIG. 95. Section of the blastoderm of an egg eighteen to twenty hours old, showing nuclei leaving the blastoderm and entering the yolk, x 840.

their formation. The immigration of secondary yolk cells from the blastoderm into the yolk is most evident in preparations of eggs eighteen to twenty hours old. It may be observed in eggs as old as twenty-four to twenty-six hours; but in these it is rare. In many preparations of eggs twenty-two to twenty-four hours old there are present scattered through the yolk, nuclei almost devoid of cytoplasm corresponding in size and general aspect to the secondary yolk cells. (Some of these may be seen in figure 16). While the identity of these cannot be demonstrated it is reasonable to suppose that they are actually the secondary yolk cells, since in their uniform size and poverty in cytoplasm they differ from the majority of the primary yolk cells. These cells,—presumably the secondary yolk cells,—are not infrequently aggregated into clusters, a few of these containing as many as ten or a dozen nuclei. A little later, in eggs twenty-eight to thirty hours old, these almost naked nuclei have disappeared. In their place are yolk cells of moderate size, resembling in every respect the smaller representatives of the primary yolk cells. It seems probable therefore that these are the secondary yolk cells which have acquired additional cytoplasm, since there is no other explanation possible of the facts observed. The means by which this increased amount of cytoplasm is acquired is not clear. It may have been manufactured by the metabolic activities of the cell, building cytoplasm out of the food substances of the yolk, or it may have been drawn directly from the protoplasmic

primary yolk cells by division, and which afterwards contributed to the formation of the fat body and the blood. Heymons (1895) states that these are nothing but blood cells, and are therefore not "yolk cells" at all.

complex within the yolk. This complex of course includes the yolk cells, the latter, as already stated, being linked together by fine processes to form a complex virtually constituting a syncitium. Moreover, although the nuclei of the yolk cells degenerate, the cytoplasm of these cells is never observed in a state of degeneration. It seems not improbable therefore that this may be transferred by streaming movements to other cells less abundantly supplied.

During Stages IV to VII the general aspect of the yolk changes but little. The cells of the cephalo-dorsal body enter the yolk at the anterior end of the egg at Stage VII, as described in a previous section. In preparations of Stage IV multinuclear yolk cells are still to be seen (Fig. 19) but they are wanting in those of later stages. Aside from the cells of the cephalo-dorsal body, the cells present in the yolk, up to Stage VII, are quite uniformly distributed and of moderate size, each containing a nucelus whose diameter seldom exceeds twice that of the nuclei of the blastoderm or its derivatives. The cytoplasm accompanying these nuclei is relatively small in quantity and lacks the dark-staining qualities and consequent opacity seen at earlier stages (see Figs. 19 and 20).

At the close of the period of the formation of the blastoderm, when the rudiments of the mesenteron and the mesoderm are beginning to show themselves (Stage IV), the cells of the blastoderm become sharply demarcated from the yolk. Up to this time the bases of the cells of the blastoderm have been intimately connected with the protoplasmic network of the yolk, and form a zone of dark-stained and much vacuolated cytoplasm, the inner cortical layer (see p. 31). This central portion of the blastoderm cells now definitely separates from the blastoderm, remaining in connection with the yolk. The inner surface of the blastoderm and the outer surface of the yolk then acquire a sharp and definite outline. The inner cortical layer now forms a thin protoplasmic pellicle around the entire yolk mass enclosing it like a sac (Figs. 18, 19, 24, 26). The peripherally situated yolk nuclei naturally enter into a close relation to the surface layer, and this relation is maintained to near the close of embryonic development. These nuclei are, however, at first few and scattered, but after the epithelium of the mid-intestine has covered the dorsal surface of the yolk they begin to appear in greater number, and at Stage XI are seen applying themselves to that portion of the peripheral layer which is in contact with the ventral margins of the epithelium of the mid-intestine, as shown in figure 75. In this figure yolk cells are also seen lying close to the dorsal surface of the yolk. Such peripheral accumulations of yolk cells are noticeable only in the anterior half of the mid-intestine. At Stage XII (Fig. 81) many of the peripheral yolk cells appear to have traveled ventrad in company with the advancing ventral margins of the epithelium of the mid-intestine and are now situated close beneath the un-covered area of the yolk, where they remain until covered over by the walls of the mid-intestine (Fig. 82). A little later these nuclei are seen to become shrunken and distorted and soon after this they disappear completely leaving no trace behind, having been digested and absorbed by the wall of the mid-intestine.

In addition to the changes described above relating principally to the yolk cells, other changes take place in the yolk between Stages IV and XIV. These concern the vitelline bodies and the protoplasmic meshwork. As early as Stage IX many of the vitelline bodies are seen to have become much enlarged. Careful scrutiny of sections under a high power shows that the vitelline bodies are undergoing a process of fusion, several joining together to form a larger body and such larger bodies in turn joining with one another to form still larger bodies. These compound bodies when first formed are lobate, the lobes corresponding to the vitelline bodies which have taken part in its formation. Later the lobes disappear and the bodies become spheroidal or ovate in form. This process of fusion goes on continuously up to Stage XIII-XIV when the vitelline bodies are but few in number but of large size, some of them attaining a diameter of eight to nine micra. Soon after this they disappear, presumably being digested, and are absent in examples of Stage XV.

At Stage IX the protoplasmic meshwork of the yolk also begins to show changes, its meshes here and there being seen to have increased in size, indicating coalescence of the vitelline spheres. This coalescence continues slowly but steadily up to Stage XIV, when the protoplasmic network of the yolk is made up of large but coarse meshes. Soon it becomes ragged, irregular and finely granular, ultimately, at Stage XV acquiring the appearance of a pale and finely granular precipitate.

The yolk cells have been given but slight consideration by most of the investigators of insect embryology. The reason for this is not far to seek. Many of the earlier embryologists looked upon the yolk cells as the material from which the mesenteron was formed, and therefore corresponding not only theoretically but actually to the entoderm of other animal forms. When, however, it was found that the mesenteron in the pterygote insects was derived from other sources, interest in the cells of the yolk waned, and consequently published observations on them are, so to speak, sporadic and scattered.

The origin of yolk cells from cleavage cells which do not migrate to the periphery of the yolk, but remain behind, was established by Bobretzsky (1878) in the case of the butterfly *Pieris*. This mode of origin of the primary yolk cells was confirmed by many other investigators. Grassi's observations on the honey bee (1884) seem to be entitled to a place among these, although his statements on this point are not perfectly clear. The origin of yolk cells in this manner has subsequently proved to be the rule among pterygote insects, although exceptions were early noted. Thus Patten (1884) in *Neophylax* and Korotneff (1885) in *Gryllotalpa,* found that all of the cleavage cells migrate to the surface of the egg. Later some of these wander back into the yolk to form the primary yolk cells. This is commonly taken to be a primitive condition, based on the assumption that the primary yolk cells represent the entoderm. This retrograde migration must then be regarded as a modified process of gastrulation like that found in arachnids.

A second characteristic which the yolk cells of the bee possess in common with certain other insects is the large size of these cells. It is certain that this is true of the Dermaptera and Orthoptera, since Heymons (1895) calls especial attention to it and it is also true, according to Lecaillon (1898) for the chrysomelid beetles *Chrysomela, Lina* and *Agelastica,* but not for *Clytra.* It seems doubtful how far this relation obtains among insects in general. Definite statements on this point are wanting and an inspection of the plates given by various investigators has proved inconclusive, although it seems probable that in most insects the primary yolk cells are at least of more than average size.

The formation of "nests" of nuclei occurs in many insects and has been especially mentioned by Heider (1889) in *Hydrophilus* and by Heymons (1895) in the Dermaptera and Orthoptera, more recently by O. Dickel (1904) and Nachtsheim (1913) in the honey bee and by Friederichs (1906) in chrysomelid beetles. Dickel and Nachtsheim state that they found mitotic figures in the yolk cells (see p. 230), but both Heymons and Friederichs found multiplication of nuclei to be due principally to amitotic (direct) division.

The degeneration of the primary yolk cells at an early stage in the development of the embryo seems to have been noted only by Lecaillon (1897) and Friederichs (1906), in the chrysomelid beetles. The latter investigator has made a special study of the cell elements of the yolk and finds that (1) the primary yolk cells multiply at first by mitosis, later by amitosis, (2) that many of their nuclei soon degenerate, forming chromatin spherules which find their way out of the cells themselves into the yolk. These he terms "entodermal paracytoids" in order to distinguish them from similar bodies formed by cells of the tissues of the embryo. As previously stated these conditions are closely approximated in the honey bee.

The yolk itself, in at least three of the large orders of insects, the Orthoptera, the Coleoptera and the Lepidoptera, becomes split up into polyhedral blocks or segments soon after the formation of the germ band. Each of these so-called segments contains one or more yolk nuclei. In the examples of the Diptera studied this process appears to be lacking, as it is also in the mason bee (Carrière and Bürger 1897), as well as in the honey bee. In the latter insect at least the absence of yolk segmentation is probably due to the fluidity of the yolk, which would make such a breaking up a physical impossibility.

The immigration of small cells or small nuclei from the blastoderm or the embryonal tissues into the yolk to form secondary yolk cells has been described by a number of investigators. Heymons (1895) and Friederichs (1906) have reviewed this subject quite completely, so that a prolonged discussion would be out of place here.

A migration of cells from the germ band into the yolk was first claimed by Graber (1871, 1878), and later (1888a) described more completely in the beetle *Melolontha*. Similar cells were also observed by Korotneff (1885) in *Gryllotalpa*, and Wheeler (1889) in *Leptinotarsa* (*Doryphora*). Voeltzkow (1889) and Noack (1901) have found that in the muscids also cells enter the yolk from the blastoderm. According to Noack some of these secondary yolk cells form true yolk cells as in the bee, while others degenerate. Mecznikow (1866) and Will (1888) mention the migration of cells into the yolk from the posterior pole of the egg in the aphids, but the relation of these to the secondary yolk cells of other insects is not clear. In the Lepidoptera, Schwarze (1899) states that in *Lasiocampa* cells detach themselves from the anterior end of the mesoderm and wander off into the yolk. Later they leave the yolk and enter the body cavity becoming blood cells. Schwangart (1904), states that the greater part of the anterior mesenteron rudiment becomes resolved into its component cells; these enter the yolk and become concerned in its absorption. Later they contribute to the formation of the mesenteron. Heymons (1895) has paid special attention to the secondary yolk cells in *Forficula, Gryllotalpa* and *Gryllus,* and which he names "paracytes." These cells are first formed from the lateral portions of the blastoderm, but may later be seen arising from all parts of the germ band, especially the mesoderm. The one unmistakable characteristic, distinguishing these cells from normal embryonic cells consists, according to Heymons, in "the dissolution of the nucleus, and particularly in an extremely characteristic separation of the chromatin from the other substances contained in the nucleus, especially the nucleolus." In other insects true "paracytes" have apparently thus far been observed only by Schwarze (1899) in *Lasiocampa* and by Friederichs (1906) in *Meloë*. The relation of the paracytes to yolk cells of secondary origin observed in other insects is at present uncertain.

Friederichs (1906) has made a especially intensive study of the cell elements of the yolk in certain beetles. Some of his conclusions regarding the primary yolk cells have already been mentioned. In the chrysomelid beetles true "paracytes" were not found. Instead, there are found issuing from the cells of the germ band—ectoderm, mesoderm, sex cells—small globular bodies formed by the emission of chromatin from certain of the nuclei of the tissues concerned. These bodies Friederichs calls "paracytoids." After entering the yolk they disappear. In appearance they resemble so closely the chromatin spherules formed by the degeneration of the nuclei of the (primary) yolk cells, that they are not distinguishable from these, and the term "paracytoid" is made to cover chromatin spherules from both sources. Both are morever supposed to be identical in function, which is presumably that of assisting in the digestion and assimilation of the yolk. In *Meloë* typical "paracytes" were found; these later degenerate in the yolk and form "paracytoids."

The aggregation of the yolk cells beneath the wall of the mid-intestine in embryos of the honey bee was noticed by Kowalevski in 1871, but this condition, as well as the occurrence of nuclei in the peripheral layer of protoplasm surrounding the yolk at earlier stages, is found in a far more pronounced form in the mason bee, *Chalicodoma* (Carrière and Bürger, 1897). Bürger's description is as follows (p. 358):

"The *yolk cells* also participate in the formation of the epithelium of the mid-intestine. Not directly indeed, in the sense of affording material for its construction, but in that they doubtless afford nourishment to the entoderm bands.

"At the time of the appearance of the entoderm bands the yolk cells at the periphery, regularly distributed and connected together, form a complete sac, or if you will, a primary mid-intestinal epithelium.

"If the series of sections, which are intended to elucidate the development of the wall bee, ceased just after the time of the breaking through of the mouth, one would be inclined to consider seriously whether the yolk cells did not form the definitive epithelium of the mid-intestine. Possessing other stages, it is however evident to us, that the nuclei of the yolk cells degenerate, forming a crumbling mass, which is absorbed together with the remaining yolk. In the oldest embryo which I have sectioned the intestine is empty of yolk and also of every trace of yolk cells and their nuclei."

This rather remarkable condition does not appear in other insects, and may be peculiar to the Hymenoptera. It recalls the condition found by Madame Tschuproff-Heymons (1899) in the Odonata and suggests that possibly the behavior of the yolk cells in the Hymenoptera may have a phylogenetic significance, the epithelium-like layer formed by the yolk cells at the periphery of the yolk representing the vestiges of the ancient mid-intestine, which undoubtedly was primitively formed by the yolk cells. If this were the case however, similar conditions should be found in the Orthoptera and other primitive orders. In *Gryllus* and *Gryllotalpa*, according to Heymons (1895) the yolk cells form an epithelial layer beneath the wall of the mid-intestine, but this does not occur until shortly prior to hatching. In *Dixippus*, according to Hammerschmidt (1910) the yolk cells form an epithelial layer over the ventral surface of the yolk at a relatively early stage. This layer Hammerschmidt regards as the "primary entoderm", a view which is in harmony with the suggestion outlined above.

XV

Duration and Rate of Development

Early in the course of the present work, efforts were made to determine the rate of development, that is, the length of time required by the egg to reach a particular stage. Unexpected difficulties were encountered, and the problem was found much less easy of successful solution than was at first anticipated. This was doubtless due to ignorance of the proper method, and much time was consumed in unsuccessful experimentation. In general two methods were tried. The first is a modification of those employed by Petrunkewitsch (1902) and Dickel (1904). It consists in removing a frame from the brood nest of a vigorous colony, and replacing it by an empty frame, upon which the queen is placed. The hive is then closed for a period of two or three hours. The experiment was varied by using two or more empty frames, examining them every two hours. This method, after many trials, was abandoned, since it was seldom successful. The failure of this series of experiments was probably due in part, at least, to the fact that they were undertaken too late in the season, after the main honey flow had ceased. Continued attempts however were made to remedy this unfavorable condition by feeding thin sugar syrup by means of a "pepper box" feeder placed above the frames. Next a single frame observation hive was used, the queen being watched continuously in the hope that some time during the day the queen would deposit during a two-hour period a number of eggs sufficient for the purpose in hand. This experiment was a total failure since not once during the several days she was under observation did she deposit more than a few eggs, and these at long and infrequent intervals, practically all of the eggs being laid during the night. It may be added that the observation hive was kept in a warm but shaded position out of doors, in one of the large breeding cages used by the entomologists of the Department. The third set of experiments was moderately successful. The method used in these was the following:

A strong colony was shaken on eight full frames of foundation and two empty combs, which were placed in the center of the hive. The two combs were after two hours usually found to contain a fair number of eggs. These combs were then marked and placed in the second story of another strong colony, above a queen excluder, combs of brood being taken from the first story and placed on each side of the combs containing eggs, in order to make sure that these would not be deserted. After the proper interval the eggs were removed from the comb, fixed, and afterwards stained and mounted for examination. This method is open to the serious objection of being very demoralizing to the colony subjected to it, and cannot safely be tried more than once on the same colony unless a considerable interval elapses between trials.

The data in regard to the rate of development, obtained by the foregoing methods is summarized in the accompanying table. This shows the approximate time in hours in the left hand column, the principal development changes observed at this age in the middle column, and the corresponding illustration in the right hand column.

Recorded data relating to this subject are scarce. The following may be gleaned from Dickel's paper (1904). At twenty hours the blastoderm is completely formed, and its nuclei arranged in a single layer (Fig. 1). According to the writer's data the nuclei do not become arranged in a single layer until a few hours later. An egg twenty-four hours old is shown in figure 2. In this the gap between the dorsal and ventral blastoderm, located just dorsad of the cephalic pole, is plainly shown, as is also the case in figure 3, which is about twenty-six hours old. In the figure last mentioned secondary yolk cells are evident, distinguishable by their lack of cytoplasm. These two figures of Dickel agree fairly well with figure 16 of the present paper, representing a sagittal section through an egg twenty-four to twenty-six hours old. Dickel's figure 10 represents an egg thirty-five hours old in which "the three germ layers are differentiated." It apparently corresponds to Stage V or VI, although the cephalic fold of the amnion is not represented. This would agree with the data in the table given below.

TABLE SHOWING RATE OF DEVELOPMENT OF EMBRYO

AGE IN HOURS	DEVELOPMENTAL CHANGES OBSERVED	FIGURES
1-6	Cleavage proceeds rapidly from one to many cells.	I and II
8-10	Cleavage cells reach the surface of the egg. [Estimated]	III
14-16	Cleavage cells cover the entire surface of the yolk. Appearance of primary yolk cells.	
18-20	Nuclei of the cells of the blastoderm become arranged in two layers. Appearance of secondary yolk cells.	
20-30	Cells of blastoderm become arranged to form a single layer of prismatic cells. Nuclei of primary yolk cells degenerating.	
32-34	Differentiation of middle plate begins.	IV
34-36	Appearance of anterior mesenteron rudiment. Cephalic fold of amnion formed.	V
36-38	The lateral folds and the posterior mesenteron rudiment are formed.	
38-42	Approximation of lateral folds. Caudal fold of amnion formed.	VI
42-44	Closure of lateral folds completed. Amnion becomes a closed sac.	VII
44-46	Formation of the stomodaeum and proctodaeum. Appearance of the rudiments of the labrum, antennae, gnathal appendages, silk gland and stigmata.	VIII
48-50	Formation of stomodaeum. The invaginations which are to form the silk gland and stigmata are still wide slits.	IX
52-54	Formation of proctodaeum and Malpighian tubules. Labrum, cerebral lobes and rudiments of the mesodermal somites now well defined. Formation of tracheal trunks and their branches. Silk glands become elongated tubules.	X
58-60	Approximation of second maxillae to form labium. [Estimated.]	XII
66-68	Second maxillae united to form labium. Heart formed. Development nearly complete.	XIII
74-76	Embryo breaks chorion and becomes a larva.	XV

Petrunkewitsch (1902) gives the following table for drone (male) eggs:

Age in hours	*Stage of development*
3-4	first cleavage spindle.
4-7	cleavage.
7-9	movement of the cleavage nuclei towards the periphery.
9-20	formation of the blastoderm.
20-25	blastoderm.
25-36	gastrulation.
36-38	end of gastrulation; rudiments of the mesodermal tubes formed.

According to these data it appears that the development of the drone egg is more rapid than that of the worker (female) egg, since "gastrulation," that is, the formation of the mesoderm and the mesenteron rudiments, commences at least five hours earlier. Unfortunately the time required for total development is not known.

Among other insects, the only data which are sufficiently complete to be of value for comparison are those of Heider (1889) for *Hydrophilus*. These are extremely full and explicit, and include all of the more important phases of development of the insect. They are arranged in the form of a table which covers two and a half folio pages and therefore far too elaborate to be reproduced here. It will nevertheless be profitable to compare *Hydrophilus* and the honey bee, using for this purpose certain more or less arbitrary periods into which the development may conveniently be divided.

Kowalevski (1871) distinguishes three periods in the development of *Hydrophilus*. They are as follows: The first extends to the complete formation of the germ band and of the embryonic envelopes. The second includes the formation of appendages and the origin of the glandular layer of the intestine ("Darmdrüsenblatt"), up to the formation of the dorsal plate and the rupture of the embryonic envelopes. The third includes the complete differentiation of the embryo and the dorsal wall of the mid-intestine. Heider (1889) interprets these periods as follows:

"In the first period are included: the formation of the embryo, of the rudiments of the cephalic lobes and of the embryonic envelopes up to their complete overgrowth of the embryo. At the end of this period the completely segmented germ band has extended its entire length over the surface of the yolk. To the second period belongs: the invaginations constituting the proctodaeum and stomodaeum, the formation of the rudiments of the appendages and of the central nervous system, and the tracheal invaginations. Into this period falls the separation of the entoderm from the mesoderm together with the first rudiments of the mid-intestine, the appearance of the primary segments, the development of the definitive body cavity and the segmentation of the yolk. The rupture of the embryonic membranes and their withdrawal to the dorsal side of the egg, and an invagination, introducing the formation of the dorsal organ, form the limits between the second and third embryonic periods. . . . No new organs are produced, this period serves principally for histological differentiation."

In dividing the development of the bee into three corresponding periods, indicated in the table given above by double lines, it is easy to determine the limit between the first and second periods, which falls between Stage VII, forty-two to forty-four hours, and VIII, forty-four to forty-six hours. The limit between the second and third periods is not so easy to find, since there is in the bee but a single embryonic membrane, which is not ruptured until hatching and which forms no "dorsal organ." Moreover in the bee egg there is no movement of the embryo corresponding to "revolution" or "blastokinesis" (Wheeler, 1893). By referring to Heider's table however, it is seen that at the end of the second period, which falls between the fourth and fifth days of development, the tracheal invaginations, formed but a few hours previously, are still cleftlike. During the fourth day moreover the invaginations of the proctodaeum and stomodaeum are formed. In the bee therefore the limit between the second and third periods has been fixed between Stages IX and X, corresponding respectively to the ages of forty-eight to fifty hours and fifty-two to fifty-four hours, which is probably sufficiently close for purposes of comparison.

In *Hydrophilus* the time required for complete development is eleven days, or 264 hours, while that of the bee is only seventy-six hours. The only way in which the length of time consumed by each period in the two insects can be compared is by expressing it as a fraction of the time required for total development. This is most conveniently expressed by decimals. The time of each period except the first, is calculated from the close of the preceding period. The result is shown in the following table:

Period	I	II	III
Hydrophilus	0.272	0.091	0.636
Apis	0.579	0.078	0.342

The discrepancy between these figures is striking. In *Hydrophilus* slightly over one-fourth of the time required for total development is taken up by the first period; in the bee over one-half is consumed. Period II consumes relatively but a trifle more time in *Hydrophilus* than in the bee, while period III consumes nearly twice as much. If the time required for the formation of the blastoderm be similarly compared, the result is expressed by the fractions 0.166 for *Hydrophilus*, and 0.39 for *Apis*. Expressed in general terms, it may be said that as compared with *Hydrophilus*, the earlier development of the bee is greatly delayed or conversely, that the later development is greatly accelerated. So little is known of the vital forces underlying development that the reason for these differences can only be surmised. It seems not improbable however that one factor is the relatively slight degree of differentiation of the bee larva, as compared with that of *Hydrophilus*. The larva of the latter insect is fitted to lead an active existence, finding, catching and killing its prey, and at the same time avoiding its enemies. It is therefore well equipped with functional organs of locomotion and perception together with a well developed central nervous system, most of which in the bee larva are present only as rudiments, and do not become functional to any considerable degree until after metamorphosis.

XVI

TECHNIQUE

Except when timed stages were desired, a frame of brood comb, containing a considerable number of eggs was selected. The eggs were then lifted out of the cells and transferred singly to the fixing fluid. A small fine-pointed brush, slightly moistened, was found most convenient for this purpose. Other investigators have made use of a bent needle, but with this instrument there is considerable danger of damaging the rather delicate egg. A considerable difference was noted between different lots of eggs in regard to the ease with which the eggs could be removed. In some lots a light touch with the brush was all that was needed to loosen the egg from its attachment to the bottom of the cell, in others the eggs were stuck so tight as to require considerable manipulation to bring them out of the cells without damage. Although no records were kept it seems not unlikely that this difference in the adhesiveness of the eggs was traceable to a difference between the mother insects, the queens, some secreting more adhesive, or else of a better quality, than others.

Five fixing fluids were used: Bouin's picro-formol, one-half to eighteen hours[25]; acetic-alcohol (Carnoy's second formula), one-half to one hour; Gilson's mixture, twenty-four hours; Petrunkewitsch's modification of Gilson's formula, twenty-four hours; and Kleinenberg's picro-sulphuric mixture, three hours. The last named was used but little. All of the others gave many satisfactory preparations and some superlatively good ones. In quality of the fixation, as regards the tissues of the embryo, there is little room for choice. The effect of the different fixing fluids on the yolk is however quite different. This has been described in the chapter on the organization of the egg (p. 12). The effect on the cells in the yolk is likewise different, eggs fixed with picro-formol showing these to the best advantage. The question as to

[25] The best material came from lots of eggs left in this for five to six hours or longer.

which fixing fluid gives the truest pictures of these cells can however only be determined by careful comparison with living eggs.

Picro-formal proved to be the best medium for fixing eggs intended to be studied entire, and since the other media showed no decided advantage in other respects, it was most frequently employed. Its reason for superiority appears to be due to the circumstance that it does not coagulate the yolk spheres, consequently the yolk remains clear and transparent. Embryos fixed in this fluid and mounted entire have a brilliancy and transparency not seen in specimens fixed in other fluids.

The eggs were in all cases preserved in eighty per cent alcohol.

Of the five fixing fluids used, Gilson's and Petrunkewitsch's fluids harden most, eggs fixed in these media becoming rather brittle, while acetic alcohol hardens the least. Eggs treated with this mixture are quite elastic and may be manipulated without much danger of breaking, although they are rather easily distorted. This is true also of eggs treated with picro-formol, although not to the same extent.

The chorion of the egg of the bee, although excessively thin, is nevertheless relatively impermeable, especially to aqueous fluids. After fixing and hardening, and before staining or clearing is attempted it is therefore necessary to puncture the chorion, which lies close to the egg except at its two poles. The chorion may be punctured or ruptured here, but the most satisfactory solution of the difficulty is to puncture the egg in one or two different places with a fine cambric needle, rubbed on an oil stone to an extremely fine point. This is best done by watching the progress of the sharpening under a binocular microscope. The punctures were usually made along the dorsal mid-line of the egg since this region is not occupied by the embryo until near the close of development.

Mayer's carmalum was used for staining eggs intended to be mounted entire. The eggs were stained from two to forty hours, and thoroughly destained in acid alcohol. This method yielded uniformly good results. Iron haematoxylin was found most satisfactory for sections, used with or without a counter stain. Mayer's paracarmine (alcoholic) was used for staining eggs preliminary to embedding.

It is absolutely necessary that clearing be accomplished gradually. If this is not done, distortion will almost inevitably result. In cases where the chorion is not opened, or opened insufficiently, a transfer from alcohol to clearing fluid will cause the egg to be completely flattened by osmotic pressure, owing to the slight resistance of the yolk. Clearing may be quickly and easily accomplished by the following method of flotation, suggested by Dr. Petrunkewitsch: A small quantity of pure cedar oil is poured into the bottom of a narrow vial. On the top of this are poured successive layers containing an increasing proportion of absolute alcohol. In practice these were: pure oil, two-thirds oil, one-third oil. On the top layer the embryos were placed surrounded by pure absolute alcohol. They then descend gradually to the bottom of the vial, when they may be taken out, washed in pure cedar oil and transferred to balsam, if they are to be mounted for study entire. If intended for sections they were placed in xylol before transfer to paraffin. Embedding was done in watch crystals. Paraffin of fifty-five degrees C. melting point was used, the eggs remaining in the paraffin from one-half to an hour to three hours.

It is extremely desirable to orient the embryo accurately. This was accomplished by placing the watch crystal containing the melted paraffin and the embryo on a piece of clear glass, slightly warmed, and placing the whole on the stage of a binocular microscope. A layer of partially cooled paraffin, sufficient to support the embryo in any position soon forms at the bottom of the watch crystal, and the embryo may be placed in this in the position desired. In practice the ventral face was placed uppermost. The watch crystal was then floated on a vessel full of cold water, and the paraffin subsequently hardened by complete immersion.

Sections were cut by means of a Minot microtome, five to seven microns in thickness, according as the conditions were favorable or otherwise.

Summary

The elongate cylindriform egg is slightly curved in the long axis and, with reference to the future embryo, presents a slightly larger cephalic end, and a convex ventral, as opposed to a concave dorsal surface. Two membranes cover the egg, an outer, the chorion, and an inner, the vitelline membrane. The chorion is extremely thin, transparent, and covered with minute papillae arranged in the form of a fine network. A thickened area at the cephalic end of the egg appears to represent a micropylar area. The vitelline membrane is extremely thin and appears to be structureless. The contents of the egg comprise a large quantity of deutoplasmic material or yolk, and a small quantity of protoplasm. The yolk is principally in the form of spherical globules, the vitelline spheres, composed of a transparent fluid of an unknown chemical nature, not fat nor oil. Within the yolk spheres are much smaller relatively solid bodies, the vitelline bodies. The protoplasm surrounds the yolk spheres, filling the interstices between them, and also forms a thin cortical layer over the surface of the yolk. Within the protoplasm are numerous minute bodies, possibly identical with the Blochmann's corpuscles of certain other insects.

The cleavage cells at first form a rounded group near the cephalic pole of the egg. They multiply rapidly and soon form an elongated hollow fusiform figure, its smaller end extending toward the caudal pole. As the cells continue to increase in number the figure also increases in size, until the cells at its larger end, on the ventral side near the cephalic pole, enter the cortical layer and begin to form blastoderm. The majority of the cleavage cells finally reach the cortical layer except a few which remain in the yolk to form the primary yolk cells.

As the cleavage cells approach the periphery of the egg their nuclei assume a peripheral position in the cells. The latter embed themselves in the cortical layer, the latter, together with the cleavage cells, constituting the blastoderm. The central ends of the cleavage cells, however, remain at first united together below the cortical layer thus forming the inner cortical layer. Later this inner cortical layer is cut off from the remainder of the cleavage cells by a structure having the appearance of a basement membrane. Meanwhile the cells of the newly formed blastoderm multiply by mitotic division, the division planes being at first normal to the surface of the egg, but later become oblique, the cells thus becoming wedge-shaped or pyriform. Finally they assume a prismatic form, the nuclei all coming to lie at the same level; the basement membrane disappears, and the greater part of the inner cortical layer is absorbed by the cells of the blastoderm, their inner ends becoming sharply defined. The remnant of the inner cortical layer clings to the yolk over which it now forms a pellicle, this being separated from the blastoderm by a narrow space. The blastoderm is at first of nearly uniform thickness around the circumference of the egg, but soon becomes differentiated into a thicker ventral and a thinner dorsal portion. Along the dorsal mid-line is a strip of cells differing from the remainder in being especially thin and flat and in maintaining a close relation with the yolk.

The mesoderm is formed from a median area of the ventral blastoderm, the middle plate, which separates from the blastoderm on each side of it and which constitutes the lateral plates. The ventral plate is then covered over by the lateral plates, which approach one another, and finally become united along the ventral mid-line to form the ectoderm. The rudiments of the mesenteron are formed by the immigration of blastoderm cells, a discoid swelling being thus produced at each of the two ends of the middle plate, but outside of its limits. These rudiments later become covered by ectoderm. The anterior mesenteron rudiment does not however become completely covered, a small circular area near its caudal margin remaining uncovered and later constituting the floor of the stomodaeal invagination. During these developmental changes the median dorsal area of the blastoderm, composed of thin flat cells, also becomes depressed and is overgrown by the dorsal margins of the lateral plates. Meanwhile the cells of this dorsal strip have becme aggregated in the cephalo-dorsal region of the egg to form a more or less discoid mass, the cephalo-dorsal body, the so-called "yolk plug" of O. Dickel (1904). During the formation of the so-called "germ-layers" both the middle and lateral plates show plain evidence of segmentation. The segments thus indicated appear to correspond to the definite segments of the embryo.

The amnion is single layered and is formed from the dorsal half of the blastoderm. This separates both from the yolk and from the margins of the ventral or embryonic half of the blastoderm, the ventral plate. This separation does not take place simultaneously, but occurs first at the cephalic end of the egg, a cap-like fold being formed which grows rapidly caudad. A similar but slighter fold is later formed at the caudal end of the egg. The two folds meet and coalesce near the caudal pole. The amnion probably corresponds to the serosa of other pterygote insects.

Twenty-one segments were found in the bee embryo, including an anal segment or telson. Appendages were observed on the antennal segment, the three gnathal segments, and the three thoracic segments. No abdominal appendages were found. The antennal rudiments and those of the three thoracic legs became reduced to hypodermal thickenings prior to hatching. The supposed appendages of the premandibular segment are only exaggerated ganglionic swellings. No evidence of the presence in the bee of a superlingual segment (Folsom) was found. As is usual in insect embryos, the second maxillae fuse to form the labrum.

The rudiments of the nervous system take the form externally of two longitudinal swellings of the ectoderm, the primitive swellings, one on each side of and close to the ventral mid-line. At the oral region these diverge, and in the cephalic region of the embryo expand into the broad procephalic lobes. The primitive swellings, including the procerebral lobes, are divided by slight intersegmental constrictions into twenty neuromeres corresponding to the segments of the embryo. Two of these, formed from the procerebral lobes, are preoral. The first, the protocerebrum, corresponds to the primary head segment, the second, the deutocerebrum, to the antennal segment. The two lobes constituting the protocerebrum become rather obscurely subdivided into three lobes, the most anterior of which become the optic lobes. During the formation of the primitive swellings the germ band lengthens, so that its cephalic end, including the procerebral lobes, becomes curved around the cephalic pole of the egg. The protocerebral lobes thus come to lie on the dorsal side of the egg with their cephalic ends directed toward its caudal pole. Although the germ band shortens somewhat at a later stage, the protocerebral lobes always remain directed caudad, and not cephalad, as in more primitive insects (*e.g.* Orthopetra). The third, neuromere, corresponding to the premandibular or intercalary segment, is postoral, since the commissure connecting its two lateral halves (ganglia) passes below the stomadaeum. These three neuromeres become united to form the supraoesophageal ganglion or brain. The neuromeres

of the three gnathal segments unite to form the suboesophageal
ganglion. The remainder, fourteen in number, become the ganglia of the ventral nerve cord. The cells of the ectoderm constituting the primitive swellings and protocerebral lobes become
segregated into two layers, an outer layer of small cells, the
dermatoblasts, destined to form hypodermis only, and an inner
layer of large cells, destined to form nerve tissue, the neuroblasts.
These divide unequally and teleblastically, giving of centrad several cells smaller than themselves. These last mentioned cells
divide equally, the products becoming differentiated to form
ganglion cells. The latter are therefore the granddaughter cells
of the neuroblasts. The optic ganglia are not however produced
by the agency of neuroblasts, but are formed as simple infoldings
of the ectoderm. Beginning at the anterior margin of the mandibular segment, and extending to the last segment of the trunk
is a narrow median strip of ectoderm, the median cord. In the
intrasegmental regions this contributes the central portions of
the ganglia, in the intersegmental regions, it constitutes a series
of thickenings of the hypodermis. The supraoesophageal commissure is formed at least in part from the median ectoderm, but
this could not be established in the case of the suboesophageal
commissure. The stomatogastric nervous system is formed from
three median evaginations of the dorsal stomodaeal wall. The
first of these furnishes a few cells contributing to the formation
of the labral nerve, the second produces the stomatogastric ganglion, and the third the pharangeal ganglia. An outer neurilemma
only is present in bee embryos. This is formed from cells having
the same origin as the ganglion cells, and which migrate to the
external surface of the brain and ventral cord. The so-called
"ganglia" or corpora allata are formed from invaginations of the
ectoderm between the bases of the rudiments of the mandibles and
first maxillae, and are closely associated with the invaginations
which produce the apodemes of the retractor muscles of the
mandibles. The invaginations producing the corpora allata soon
become solid masses of cells, lose their connection with the external ectoderm and migrate mesiad and dorsad, finally becoming
attached to the ventro-lateral angles of the coelomic sacs of the
antennal segment. Degenerating cells are found in the rudiments

of the brain of all embryos, being especially abundant in the region between the second and third lobes of the protocerebrum. They appear soon after the brain begins to form and are evident up to the time of hatching of the embryo. Their significance is not known.

The tracheal system is formed from eleven pairs of invaginations of the lateral ectoderm. The first of these, situated on the second maxillary segment, by the formation of four diverticula, produces the anterior ends of the main tracheal trunks, including the anterior tracheal commissure or loop, and also the tracheae supplying the head. The ten pairs remaining are situated on the second and third thoracic, and the first eight abdominal segments. These also form each four diverticula. The anterior and posterior diverticula become united along each side of the embryo, thus forming the longitudinal tracheal trunks, the ventral diverticula fuse with those of the opposite side of the same segments to form the tracheal commissures, the dorsal diverticula form branches supplying the dorsal region of the larva. The openings of the tracheal invaginations remain as the spiracles.

The tentorium is formed from two pairs of ectodermal invaginations. The first pair of these is situated in front of the bases of the mandibles, the second behind the bases of the first maxillae. The invaginations belonging to the first pair grow caudad and mesiad, those belonging to the second pair cephalad and mesiad. All four meet in the median plane to form a structure having the form of an X, extending across the head capsule between the oesophagus and the suboesophageal ganglion. An invagination situated immediately caudad of the base of the rudiments of the mandible produces the apodeme for the adductor muscle of the mandible.

The silk glands appear at an early period as tubular invaginations of the ectoderm caudad and mesiad of the bases of the second maxillae. These invaginations rapidly lengthen to form long slender tubes extending the length of the trunk. As the second maxillae approach one another, during the later stages of development, they carry with them the openings of the silk glands, which are thus brought to the ventral mid-line. The second maxillae are then united in such a way that a median unpaired duct is formed by them for the silk glands, opening near the tip of the labrum.

The oenocytes are produced by immigration of cells from localized areas of the lateral ectoderm. There are eight pairs of these situated on the first eight abdominal segments, in line with the openings of the tracheal invaginations.

The mesoderm, soon after its formation, becomes differentiated laterally into two layers, an outer somatic and an inner splanchnic layer, while along the ventral mid-line it remains single layered. Separate coelomic sacs are not present, the somatic and splanchnic layers of each side being continuous longitudinally throughout the trunk. At the lateral margins of the mesoderm the two layers are continuous and in this region are composed of long columnar cells. The fate of the various parts of the mesoderm is as follows: The median single-layered section breaks up into the rounded blood cells. The somatic layer forms the trunk muscles, both longitudinal and oblique, the pericardial fat cells, and the dorsal diaphragm, including the pericardial cells. The splanchnic layer sends off from its dorsal border a mesial layer which forms the muscular layer of the mid-intestine. The remainder of this layer is principally concerned in the formation of the two main divisions of the fat body. The heart is formed by the union, along the dorsal mid-line, of two rows of cells, the cardioblasts, which are derived from the angle formed by the union of the somatic and splanchnic layers at the lateral margins of the mesoderm. A mass of mesoderm cells, forming the anterior end of the mesoderm, and evidently belonging to the primary head segment, closely surrounds the stomodaeum at its appearance, and later forms the muscular layer of the fore intestine. A similar mass at the posterior end of the embryo forms the muscular layer of the hind-intestine.

The ovaries are apparently derived from the cells' constituting the genital ridges; these cells are not visually distinguishable from undifferentiated mesoderm cells. The two genital ridges are formed from the dorsal portion of the splanchnic layer, in the fifth to the tenth abdominal segments inclusive. This portion becomes detached from the remainder of the splanchnic layer. During the development of the embryo the genital ridge gradually shortens, finally occupying a position in the seventh to the ninth segments inclusive. Meanwhile it loses its attachment to the dorsal splanchnic layer, at the same time receiving an investment of cells from the splanchnic mesoderm lying immediately ventrad of it. This investiture, composed of flat cells then, contracts an adhesion with the ventral border of the heart.

The mid-intestine is formed from the anterior and posterior mesenteron rudiments in the following manner. The discoid anterior mesenteron rudiment becomes transferred from the ventral to the dorsal side of the cephalic pole of the egg by a lengthening of the embryo. At the same time the rudiment increases in superficial area, covering the cephalic end of the yolk like a cap. Its caudal margin now extends rapidly caudad over the dorsal surface of the yolk. Meanwhile the posterior mesenteron rudiment has been similarly transferred to the dorsal side of the caudal end of the egg. It now sends out a thin tongue-like process cephalad over the dorsal surface of the yolk. The caudad extension of the anterior mesenteron rudiment, and the cephalad extension of the posterior mesenteron rudiment next meet on the dorsal surface of the yolk about one-third of the length of the egg from its cephalic pole. The epithelial strip thus formed extends rapidly ventrad over the sides of the yolk until the latter is completely enclosed, the two margins of the epithelium meeting and uniting along the ventral mid-line of the yolk a short time before hatching. Both the fore and hind intestines are formed, as is usual, by ectodermal invaginations. The invagination for the fore intestine, the stomadaeum, is however, not completely ectodermal, since its floor is formed by cells belonging to the anterior mesenteron rudiment, which is not covered by ectoderm. The hind intestine is exclusively ectodermal. The lumen of the stomodaeum (oesophagus) becomes connected with that of the mid-intestine shortly before hatching. A proventricular valve is also formed at this time by folding of the stomodaeal or oesophageal wall. The lumen of the proctodaeum (hind-intestine) is at no time in connection with that of the mid-intestine, both the cephalic end of the hind intestine and the caudal end of the mid-intestine being blind. The four Malpighian tubules are formed as ectodermal invaginations which make their appearance prior to the formation of the protodaeum, grouped around the point where the proctodaeum is to appear. Four separate invaginations have not been observed, the pair situated on each side of the mid-line being connected by a shallow crescentic groove.

The primary yolk cells, which are derived from cleavage cells remaining within the yolk, multiply by mitosis, the mitotic figures being at first similar to those of the cleavage cells. A little later irregular mitotic figures are found. These are usually minute, and have the appearance of being in some cases multipolar, and in others unequal. Multinucleate cells soon become abundant, and some of these cells are of large size. Degeneration of the nuclei of the yolk cells soon becomes frequent, such nuclei diminishing in size and finally becoming reduced to minute deep-staining spherules which leave the cell body and enter the yolk. Secondary yolk cells are formed by the immigration of nuclei from the blastoderm into the yolk. These soon become indistinguishable from the primary yolk cells. Yolk cells are found distributed through the yolk until shortly before hatching. They are frequently seen clustered under the epithelium of the mid-intestine, during the time when the latter is engaged in covering the lateral faces of the yolk. Yolk cells and yolk disintegrate at the time of hatching, being presumably digested.

The total time normally required for the development of the egg is seventy-six hours. This is divided approximately as follows: Cleavage, fourteen to sixteen hours; formation of the blastoderm, fourteen to sixteen hours; formation of mesoderm, rudiments of mesenteron and embryonic envelope, twelve to fourteen hours; remainder of development, including differentiation of tissues and organs, thirty-two or thirty-four hours. The earlier stages, including the formation of the so-called "germ layers," the amnion, etc., occupy considerably over one-half of the time required for total development.

LIST OF ABBREVIATIONS

In making up this list, the abbreviations used by Snodgrass (1910) were retained to indicate the corresponding part or its rudiment in the embryo. In every case the symbols were designed to suggest the name of the part, in order to avoid frequent reference to the subjoined list.

Am, amnion.
1am, cephalic fold of amnion.
2am, caudal fold of amnion.
Ant, antenna.
AMR, anterior mesenteron rudiment.
AntL, antennal lobe (of brain).
AntMeso, mesoderm of antennal somite.
AntNv, antennal nerve.
AntR, antennal rudiment.
Ao, aorta.
ATraL, anterior tracheal loop.
Bl, blastoderm.
BlC, blood corpuscles.
1Br, protocerebrum.
1Br₁, 1Br₂, 1Br₃, the three lobes of the protocerebrum.
2Br, deutocerebrum.
3Br, tritocerebrum.
CB, cephalo-dorsal body.
Cbl, cardioblasts.
CC, cleavage cells.
Com, commissure.
Con, connective (of ventral nerve cord).
CorAll, corpora (ganglia) allata.
CL, cortical protoplasmic layer.
Ctl, cuticle.
Dbl, dermatoblasts.
DDph, dorsal diaphragm.
DegCl, degenerating cells.
DLMcl, dorsal longitudinal muscles.
DphCl, cells of dorsal diaphragm.
Ds, dorsal strip (of blastoderm).
Ect, ectoderm.
FInt, fore intestine (oesophagus).
FtCom, frontal commissure.

FtGng, frontal ganglion.

FtNv, frontal nerve.

Gl, *gland*.

GngC, ganglion cells.

Ht, heart.

HtCl, pericardial cells.

HInt, hind intestine.

Hyp, hypodermis.

ICL, inner cortical layer.

1L, *2L*, *3L*, legs (rudiments).

Lb, labium.

LC, lateral cords.

LF, lateral folds.

Lm, labrum.

LNv, lateral nerve.

LP, lateral plate.

LTraT, longitundinal tracheal trunk.
M, micropylar area.

Mal, Malpighian tubules.

MC, median cord.

MclEnt, muscular layer of alimentary canal.

Md, mandible.

MdNv, mandibular nerve.

Meso, mesoderm.

Meso 1-3L, mesoderm of leg rudiments.

MInt, mid-intestine.

MP, middle plate.

Mth, mouth.

1Mx, first maxilla.

2Mx, second maxilla.

Nbl, neuroblasts.

NlG, neural groove.

Nlm, neurolemma.

NlR, neural ridges.

Nv, nerve.

NvF, nerve fibres, "punksubstanz."

Oe, oesophagus, fore intestine.

OeCom, circumoesophageal commmissure.

Oen, oenocytes.

OMcl, oblique trunk muscles.

OpL, optic lobe, equivalent to the first lobe of the protocerebrum $1Br_1$.

OpPl, optic plate.

Ost, ostium.

Ov, ovary (rudiment).

ParC, paracardial cells.

Phy, pharynx.

PhyGng, pharyngeal ganglia.

PMR, posterior rudiment of mid-intestine.

PP, polar protoplasm.

Pro, Proc, proctodaeum.

ProL, procephalic lobes.

PriSw, primitive swellings.

PTraL, posterior tracheal loop.

RAp, apodeme of flexor (adductor) muscle of mandible.

RMcl, flexor (adductor) muscle of mandible.

SlkD, common duct of silk glands.

SlkDo, opening of duct of silk glands.

SlkGl, silk gland.

SocCom, suboesophageal commissure.

SocGl, suboesophageal ganglion.

Sp, spiracle.

SpBr, spiracular branch.

StgNv, stomatogastric nerve.

StgR, rudiment of stomatogastric nervous system.

Sto, Stom, stomadeum.

SupCom, supraoesophageal commissure.

Tae, taenidia.

Ten, central body of tentorium.

1Ten, 2Ten, anterior and posterior arms of tentorium.

Tra, trachea.

TraBr, tracheal branch.

TraCom, tracheal commissure.

TraInv, tracheal invagination.

TraTr, tracheal trunk.

VDph, ventral diaphragm.

VLMcl, ventral longitudinal muscles.

VNC, ventral nerve cord.

x, posterior tier of cells of optic ganglion.

Y, yolk.

YC, yolk cells.

y, subspherical group of cells in lateral region of deutocerebrum.

BIBLIOGRAPHY

ANGLAŜ, J. 1901. Observations sur les métamorphoses internes de la guepe et de l'abeille. Bul. Sci. France et Belg. XXXIV, pp. 363-473. Planches XIX á XXIII et 8 fig. dans le texte.

AYERS, HOWARD. 1884. On the development of *Oecanthus niveus* and its parasite, *Teleas*. Mem. Boston Soc. Nat. Hist. III, pp. 225-281, Plates 18-25.

BALBIANI, E. G. 1882. Sur la signification des cellules polaires des insectes. C.R. Acad. Sci. [Paris] XCV, pp. 927-929.

——————— 1885. Contribution à l'étude de la formation des organes sexuels chez les insectes. Rec. Zool. Suisse, II. pp. 527-588, Planches XVI bis et XVII.

BALFOUR, FRANCIS M., M.A., F.R.S. 1880-1881. A treatise on comparative embryology, London, Vol. I, 1880, Vol. II, 1881.

BLOCHMANN, F. 1884. Ueber die Reifung der Eier bei Ameisen und Wespen. Festschrift zur Feier des Fünfhundertjährigen Bestehens der Ruperto-Carola, dargebracht von dem Naturhistorisch A Medicinischen Verein zu Heidelburg, pp. 143-172. Mit einem Doppeltafel.

——————— 1886. Ueber eine Metamorphose der Kerne in den Ovarialeirn und über den Beginn der Blastodermbildung bei den Ameisen. Verhandl. Naturhist-Med. Ver. Heidelberg. pp. 243-247.

——————— 1887. Über das regelmässige Vorkommen von bacterienähnlichen Gebilden in den Geweben und den Eiern verschiedener Insecten. Ztschr. Biol. XXIV, pp. 1-15. Mit Tafel I.

——————— 1887a. Ueber die Richtungskörper bei Insekteneiern. Biol. Centbl. VII, pp. 108-111.

——————— 1889. Ueber die Zahl der Richtungskörper bei befruchteten und unbefruchteten Bieneneiern. Morph. Jahrb. XV, 1, pp. 85-96, mit Tafel V.

——————— 1892. Ueber das Vorkommen von bakterienähnlichen Gebilden in den Geweben und Eiern verschiedener Insekten. Centbl. Bakt. Parasit. XI, pp. 234-240.

BOBRETZKY, N. 1878. Ueber die Bildung der Blastoderms und der Keimblätter bei den Insecten. Ztschr. Wiss. Zool. pp. 195-215, mit Tafel XIV.

BOVERI, TH. 1887. Ueber Differenzierung der Zellkerne während der Furchung des Eies von *Ascaris meg.* Anat. Anz. II, 22, pp. 688-693.

BRANDT, ALEXANDER JUN. 1869. Beiträge zur Entwicklungsgeschichte der Libelluliden und Hemipteren, mit besonderer Berücksichtigung der Embryonalhülle derselben. Mém. Acad. Imp. St. Petersb. VIIᵉ sér., XIII, 1. pp. 1-33. Mit 3 Tafeln.

BRUCE, A.T. 1887. Observations on the embryology of insects and arachnids. Baltimore. 31 pp. 6 plates.

BÜTSCHLI, O. 1870. Zur Entwicklungsgeschichte der Biene. Ztschr. Wiss. Zool. XX, 4, pp. 519-564. Mit Tafel XXIV-XXVII.

——————— 1888. Bemerkungen über die Entwicklungsgeschichte von *Musca*. Morph. Jahrb. XIV. 1. pp. 170-174. 3 Abbildungen im Text.

CARRIÈRE, JUSTUS. 1890. Die Entwicklung der Mauerbiene (*Chalicodoma muraria* Fabr.) im Ei. Arch. Mikros. Anat. XXXV, 4, pp. 141-165. Tafel VIII.

——————— und BÜRGER, OTTO. 1897. Die Entwicklungsgeschichte der Mauerbiene (*Chalicodoma muraria* Fahr.) im Ei. Abhandl. K. Leopold-Carolin. Deut. Akad. Naturf. LXIX, 2, pp. 253-420. Mit 13 Tafel.

CHOLODKOVSKY, N. A. 1888. Ueber die Bildung des Entoderms bei *Blatta germanica*. Zool. Anz. XI, 275, pp. 163-166. 2 Figg. im Text.

———— 1889. Studien zur Entwicklungsgeschichte der Insecten. Ztschr. Wiss. Zool. XLVIII, 1, pp. 89-100, Mit Tafel VIII.

———— 1891. Ueber die Entwicklung des centralen Nervensystems bei *Blatta germanica*. Zool. Anz. XIV, pp. 115-116.

———— 1891a. Ueber einige Formen des Blastopors bei mesoblastischen Eiern. Zool. Anz. XIV, pp. 159-160.

———— 1891b. Zur Embryologie der Insecten. Zool. Anz. XIV, pp. 465-470.

———— 1891c. Die Embryonalentwicklung von *Phyllodromia* (*Blatta*) *germanica*. Mem. Acad. Imp. Sci. St. Petersb. XXXVIII, 5, pp. 1-120. Mit 6 Tafeln.

CLAYPOLE, AGNES MARY. 1898. The embryology and oögenesis of *Anurida maritima* (Guér.) Jour. Morph. XIV, 2, pp. 219-300, pls. XX-XXV.

COMSTOCK, JOHN HENRY, and KOCHI, CHUJIRO. 1902. The skeleton of the head of insects. Amer. Nat. XXXVI, 421, pp. 13-45. Figs. 1-29 in the text.

CONKLIN, EDWIN GRANT. 1897. The Embryology of Crepidula. A contribution to the cell lineage and early development of some marine gasteropods. Journ. Morph. XIII, 1, pp. 1-205. With 13 text figures and 9 plates.

CZERSKI, S. 1910. Die Entwicklelung der Mitteldarmanlage bei *Meloë violaceus* Marsch. Arch. Polon. Biol. Med. II. pp. 259-284. Taf. VIII.

DEEGENER, PAUL. 1900. Entwicklung der Mundwerkzeuge und des Darmkanals von *Hydrophilus*. Ztschr. Wiss. Zool., LXVIII, 1, pp. 113-168. Mit Tafel VIII-X.

DICKEL, OTTO. 1904. Entwicklungsgeschichtliche Studien am Bienenei. Ztschr. Wiss. Zool. LXXVII, 3, pp. 481-527. Mit Tafel XIX und XX und 46 Figuren im Text.

DOHRN, ANTON. 1866. Zur Embryologie der Arthropoden. Centbl. Med. Wiss., LIV, pp. 849-851. Berlin.

———— 1876. Notizen zur Kenntniss der Insectenentwicklung. Ztschr. Wiss. Zool. XXVI, 1, pp. 112-138.

ESCHERISCH, K. 1900. Ueber die Bildung der Keimblätter bei den Musciden. Abhandl. K. Leopold-Carolin. Deut. Akad. Naturf. LXXVII, 4, pp. 303-364. Mit 3 Doppeltafeln und 10 Figuren im Text.

———— 1900a. Ueber die Keimblätterbildung bei Musciden. Verhandl. Deut. Zool. Gesell. pp. 130-134.

———— 1901. Das Insekten-Entoderm. Ein Beitrag zur Keimblätterlehre. Biol. Centbl. XXI. pp. 416-431.

———— 1902. Ueber den sogen. "Mittelstrang" der Insekten. Biol. Centbl. XXII, 6, pp. 179-181. Mit 4 Figuren in Text.

———— 1902a. Zur Entwicklung des Nervensystem der Musciden, mit besonderer Berücksichtigung des sog. Mittelstanges. Ztschr. Wiss. Zool. LXXI, 4, pp. 525-549, mit Tafel XXX.

FOLSOM, J. W. 1900. The development of the mouth-parts of *Anurida maritima* Guér. Bul. Mus. Harvard Col. XXXVI, 5, pp. 87-157. Pls. I-VIII.

FRIEDERICHS, KARL. 1906. Untersuchungen über die Entstehung der Keimblätter und Bildung des Mitteldarns bei Käfern. Nova Acta. Abhandl. K. Leopold-Carolin. Deut. Akad. Naturf. LXXXV. Nr. 3, pp. [265]-382. Mit 7 Tafeln Nr. XXV-XXXI und 26 Fguren im Text.

Ganin, M. 1869. Ueber die Embryonalhülle der Hymenopteren-und Lepidopterenembryonen. Mém. Acad. St. Pétersb. (7). XIV. 5, Mit 1 Tafel.

———— 1869a. Beiträge zur Erkenntniss der Entwicklungsgeschichte bei den Insecten. Ztschr. Wiss. Zool. XIX, 3, pp. 381-451, mit Tafel XXX-XXXIII.

———— 1874. Ueber das Darmdrüsenblatt der Arthropoden. Warschauer Universitätsberichte.

Glaser, R. W. 1912. A contribution to our knowledge of the function of the oenocytes of insects. Biol. Bull. XXIII, 4, pp. 213-224.

Graber, Veit. 1871. Vergleichende Lebens-und Entwicklungsgeschichte der Insecten. Naturkräfte, XXII, München bei Oldenburg.

———— 1878. Vorläufige Ergebnisse einer grösseren Arbeit uber vergleichende Embryologie der Insekten. Arch. Mikros. Anat. XV. 4, pp. 630-640.

————. 1888. Ueber die primare Segmentirung des Keimstreifs der Insekten. Morph. Jahrb. XIV, pp. 345-368, Mit Tafel XIV-XV und 4 Holzschnitten.

———— 1888a. Vergleichende Studien über die Keimhüllen und die Ruckenbildung der Insekten. Denkschr. Math-Naturwiss. Kl. Kais. Akad. Wiss. Wien. LV, 2. pp. 109-162. Mit 8 Tafeln and 32 Textfiguren.

———— 1889. Vergleichende Studien über die Embryologie der Insekten und insbesondere der Musciden. Denkschr. Math-Naturwiss. Kl. Kais. Akad. Wiss. Wien. LVI, pp. 257-314. Mit 10 colorirten Tafeln und 12 Text Figuren.

———— 1890. Vergleichende Studien am Keimstreifen der Insekten. Denkschr, Math.-Naturwiss. Kl. Kais. Akad. Wiss. Wien. LVII, pp. 621-734.Mit 12 colorirten Tafeln und 38 Textfiguren.

———— 1891. Beitrage zur vergleichenden Embryologie der Insekten. Denksch. Math.-Naturwiss. Kl. Kais. Akad. Wiss. Wien. LVIII, pp. 803-866. Mit 7 Tafeln und 12 Text figuren.

———— 1891a. Ueber die embryonale Anlage des Blut-und Fettgewebes der Insecten. Biol. Centbl. XI, 7-8, Leipzig, pp. 212-224.

———— 1891b. Zur Embryologie der Insecten. Zool. Anz. XIV, pp. 286-291. Summary in Jour. Roy Micr. Soc. 1891, p. 729.

Grassi, Battista, Dr. Prof. 1882-4. Studi sugli artropodi. Intorno allo sviluppo dell Api nell' uovo. (Letta nella seduta ordinaria del 2 Marzo 1884). Atti dell' Academie Gioenia di scienze naturali in Catania. Serie 3, Vol. XVIII. pp. 154-222. Tav. I-X.

Häcker, Valentine. 1897. Die Keimbahn von Cyclops. Arch. Mikros. Anat. XLIX, 1, pp. 35-91. Mit Tafel IV und V und 5 Figuren im Text.

Hallez, Paul. 1886. Loi de l'orientation de l'embryon chez les Insectes. C. R. Acad. Sci. [Paris]. CIII, pp. 606-608.

Hammerschmidt, Johann. 1910. Beiträge zur Entwicklung der Phasmatiden. Ztschr. Wiss. Zool. LXLV, 2, pp. 221-242. Mit Tafel IV und V.

Hatschek, Berthold. 1877. Beiträge zur Entwicklungsgeschichte der Lepidopteren. Jen. Ztschr. Naturw. XI. (4) pp. 115-145. Tafel VII-IX, 2 Figg. im Text.

Hegner, R. W. 1908. Effects of removing the germ-cell determinants from the eggs of some chrysomelid beetles. Preliminary report. Biol. Bul. XVI, 1, pp. 19-26.

———— 1914. Studies on germ cells. I. The history of the germ cells in insects with special reference to the keimbahn determinants. II. The origin and significance of the keimbahn-determinants in animals.

Journ. Morph. XXV, 3, pp. 375-509. With 26 text figures and 10 single plates.

HEIDER, KARL. 1885. Ueber die Anlage der Keimblätter von *Hydrophilus piceus* L. Abhandl. Kgl. Preuss. Acad. Wiss. Berlin. pp. 1-47, Tafel I-II.

———— 1889. Die Entwicklung von *Hydrophilus piceus* L. Erster Theil. Herausgegeben mit Unterstützung der Kgl. Acad. d. Wissensch. zu Berlin. Jena. Gustav Fischer.

HENKING, H. 1892. Untersuchungen über die ersten Entwicklungsvorgänge in den Eiern der Insekten. III. Specielles und Allegemeines. Ztschr. Wiss. Zool. LIV, pp. 1-274, mit Tafel I-XII, und 12 Holzschnitten.

HENNEGUY, L[OUIS] FÉLIX. 1904. Les insectes. Morphologie—reproduction—embryogenie. Lecons recueillies par A. Lécaillon and G. Poirault. Avec. 622 figures en noir et 4 planches en coulour hors texte. Paris, Masson et cie, editeurs. pp. XX-804.

HERTWIG, R. 1881. Anlage der Keimblätter bei den Insecten. Jen. Ztschr. Naturw. XIV. Supp. 1, pp. 124-128.

HEYMONS, RICHARD. 1894. Ueber die Bildung der Keimblätter bei den Insekten. Sitzber. Akad. Wiss. Berlin. 1894, I. pp. 23-27.

———— 1895. Die Embryonalentwicklung von Dermapteren und Orthopteren unter besonderer Berücksichtigung der Keimblätterbildung. Monographisch bearbeitet. Mit 12 lithographischen Tafeln und 33 Abbildungen im Text. Jena. Verlag von Gustav Fischer.

———— 1895a. Die Segmentirung des Insectenkörpers. Anhang zu Abhandl. Akad. Wiss. Berlin, pp. 1-39, mit einem Tafel. Summary by Verhoeff, in Zool. Zentbl. II, p. 592.

———— 1897. Entwicklungsgeschichtliche Untersuchungen an *Lepisma saccharina* L. Ztschr. Wiss. Zool. LXII, 4, pp. 583-631.

———— 1897a. Ueber die Organisation und Entwicklung von *Bacillus rossi* Fabr. Sitzber. Akad. Wiss. Berlin. pp. 363-373. Mit einem Fig. im Text.

———— 1897b. Ueber Bildung und Bau des Darmkanals bei niedern Insekten. Sitzber. Gesell. Naturf. Freunde Berlin. pp. 111-119.

———— 1899. Ueber bläschenformige Organe bei den Gespenstheuschrecken. Ein Beiträg zur Kenntniss des Eingeweidenervensystems bei den Insecten. Sitzber. K. Preuss. Akad. Wiss. (Berlin), 1899. pp. 563-575. Mit 2 Figuren im Text.

HIRSCHLER, M. J. 1905. Badania embryologiczne nad motylem *Catocala nupta* L. (Lepidoptera) [Embryological studies on *Catocala nupta* L. Bull. Internat. Acad. Sci. Krakow. 10, pp. 802-810.

HIRSCHLER, JAN. 1909. Die Embryonalentwicklung von *Donacia crassipes* L. Ztschr. Wiss. Zool. XCII, 4, pp. 627-744. Mit 15 Figuren im Text und 5 Tafeln.

———— 1909a. O rozwoju listkow zarodkowych i jelita u *Gasteroidea viridula* Deg. (Coleoptera). [On the development of the germ layers and the alimentary canal of *Gasteroidea viridula* Deg. (Coleoptera)]. Bull. Internat. Acad. Sci. Krakow. 1909, 2, pp. 284-308. (Planche IV).

JANET, C. 1899. Essai sur la constitution morphologique de la tete de l'Insecte. Paris. 74 pp., 7 pls. Review by Heymons, Zool. Centralbl. VII, pp. 800-804.

———— 1899a. Sur les nerfs céphaliques, les corpora allata et le tentorium de la fourmi (*Myrmica rubra* L.) Etudes sur les fourmis, les guepes et les abeilles. Note 20. Mém. Soc. Zool. de France. XII, pp. 295-336. Planches III-VI. 3 fig. dans le text.

———— 1900. Recherches sur l'anatomie de la fourmi et essai sur la

constitution morphologique de la téte de insecte. Con planches 1
à 15, et 50 figs. dans le text. Paris, 8vo. pp. 1-205.

———— 1907. Histolyse des muscles de mise en place des ailes, après le
vol nuptial, chez les fourmis. C. R. Acad. Sci. [Paris]) 145, pp. 1205-
1208.

KARAWAIEW, W. 1893. [On the embryonic development of *Pyrrchocoris
apterus*.] Zapiski Kiev. Obsch. XIII, pp. 1-34. [Russian].

KOROTNEFF, A. 1885. Die Embryologie der Gryllotalpa. Ztschr. Wiss.
Zool. XLI, 4, Taf. XXIX-XXXI.

———— 1894. Zur Entwicklung des Mitteldarms bei den Arthropoden.
Biol. Centlbl. XIV, 12, pp. 433-434.

KORSCHELT, E. und HEIDER, K. 1891-1893. Lehrbuch der vergleichenden
Entwicklungsgeschichte der wirbellosen Thiere. Jena.

———— 1895-1900. An English translation under the title "Text-book
of the embryology of invertebrates," London.

———— 1910. Allgemeiner Theil. Vierte Lieferung. VIII. Kapital. Keimblätterbildung. E. Insekten. pp. 380-416. Mit Figuren 266-287.

KOSCHEVNIKOV, G. A. Ueber den Fettkörper und die Oenocyten der Honigbiene (*Apis mellifera* L.) Zool. Anz. XXIII, 337-353. Summary by
Heymons, Zool. Centlbl. VIII, pp. 172-173.

———— 1900. [Contributions to the natural history of the bee.] Zool.
Div. Imp. Soc. Nat. Hist. Anthrop. and Ethnog. XCIX. Contrib.
Zool. Div. XIV, pp. 1-144. With 37 text figures and 3 plates. [Russian.[

KOWALEVSKI, A. 1871. Embryologische Studien an Würmern und Arthropoden. Mém. Acad. Impér. Sci. St. Petersbourg (7) XVI, 12. 1-70.
Mit 12 Tafeln.

KOWALEVSKY, A. 1886. Zur embryonalen Entwicklung der Musciden. Biol.
Centlbl. VI, pp. 49-54. [Summary in Jour. Roy. Micr. Soc. (2) VI, pp.
429-430.]

KULAGIN, NICOLAUS. 1892. Zur Entwicklungsgeschichte der parasitischen
Hautflügler. Zool. Anz. XV, pp. 85-87. Summary in J. R. Micr. Soc.
1892, 470.

———— 1897. Beiträge zur Kenntniss der Entwicklungsgeschichte von
Platygaster. Ztschr. Wiss. Zool. LXIII. 2, pp. 195-235. Mit Tafel X
und XI.

LÉCAILLON, A. 1897. Note preliminaire relative aux feuillets germanitifs
des Coléoptères. C. R. Soc. Biol. [Paris]. pp. 1014-1016.

———— 1897a. Contribution a l'études des premiers phénomènes du
développement embryonnaire chez les insectes particulièrement chez
les Coléoptères. Arch. Anat. Micros. I, pp. 205-224. Planche XI.

LÉCAILLON, A. 1898. Recherches sur le développement embryonnaire de
quelques chrysomélides. Arch. Anat. Micros. II, pp. 118-176. Planche
VI; pp. 189-248, planche IX. Reviewed by Heymons. Zool. Zentbl.
V, 1898, pp. 813-816.

LEUCKART, R. 1865. Die ungeschlechtliche Fortpflanzung der Cecidomy-
ienlarven. Arch. Naturgesch. XXXI, 1 X, pp. 286-303, Hierzu Tafel
XII.

MARSHALL, W. S., and DERNEHL, PAUL H. 1905. Contributions toward
the embryology of *Polistes pallipes* (Hymenopteron). I. The formation of the blastoderm and the first arrangement of its cells. Ztschr.
Wiss. Zool. LXXX, 1, pp. 122-154.

MAYER, PAUL. 1876. Ueber Ontogenie und Phylogenie der Insecten. Jen
Ztschr. Naturwiss. X. (III) pp. 125-220. Hierzu Tafel VI u. VI
a, b, c.

MECZNIKOFF, CAND. [ELIAS.] 1865. Ueber die Entwicklung der Cecidomyienlarve aus dem Pseudovum. Vorläufige Mittheilung. Arch. Naturgesch. XXXI, 1, pp. 304-310.

Mecznikow, Elias. 1866. Embryologische studien an Insekten. Ztschr. Wiss. Zool. XVI, 4, pp. 389-500. Mit Tafel XXIII-XXX.

Mercier, L. 1906. Les corps bactériodes de la blatte (*Periplaneta orientalis*): *Bacillus Cuenoti* (n. sp. L. Mercier). (Note preliminaire). C. R. Soc. Biol. Paris LXI, pp. 682-684.

Nachtsheim, Hans. 1913. Cytologische Studien über die Geschlechts-bestimmung bei der Honigbiene (*Apis mellifica* L.) Arch. Zellforsch. XI, 2, pp. 169-241. Mit 6 Figuren im Text, 4 Tafeln und einer Tabella.

Nelson, Jas. A. 1911. The origin of the rudiments of the mesenteron in the honey bee. Science, N. S. XXXIII, 842, pp. 273-274.

———— 1912. A peculiar structure in the embryo of the honey bee. Science, N. S. XXXV, 899, p. 475.

———— 1914. A pair of tracheal invaginations in the second maxillary segment of the embryo of the honey bee. Science, N. S. XXXIX, 1003, p. 437.

Noack, W. 1901. Beiträge zur Entwicklungsgeschichte der Musciden. Ztschr. Wiss. Zool. LXX, 1, pp. 1-56. Mit Tafel I-V und 10 Figuren im Text.

Nusbaum, Josef. 1883. Vorläufige Mitteilung über die Chorda der Arthropoden. Zool. Anz. VI.

———— 1886. The embryonic development of the cockroach. In Miall and Denny, "The structure and life-history of the cockroach (*Periplaneta orientalis*)," pp. 181-204 with figures 104-118.

———— 1888. Die Entwicklung der Keimblätter bei *Meloë proscarabaeus*, Marsham. Biol. Centbl. VIII, 15, pp. 499-452. 2 Figg. im Text Summary, J. R. Micr. Soc. 1887, p. 942.

———— 1890. Zur Frage der Rückenbildung bei den Insectenembryonen. Biol. Centbl. X. 4, pp. 110-114.

Nusbaum, Josef, und Fulinski, Benedykt. 1906. Uber die Bildung der Mitteldarmanlage bei *Phyllodromia* (*Blatta*) *germanica* L. Mit 15 Figuren. Zool. Anz. XXX, pp. 362-381.

———— 1909. Zur Entwicklungsgeschichte des Darmdrüsenblattes bei *Gryllotalpa vulgaris* Latr. Ztschr. Wiss. Zool. XCIII, 2, pp. 306-348. Mit 2 Tafeln und 5 Figuren im Text.

Palmén, J. A. 1877. Zur Morphologie des Tracheensystems. Leipzig.

Patten, William, 1884. The development of Phryganids, with a preliminary note on the development of *Blatta germanica*.. Quart. Journ. Micr. Sc. XXIV, pp. 549-602. Pl. XXXVIa, XXXVIb, and XXXVIc.

———— 1887. Studies on the eyes of arthropods. Development of the eyes of *Vespa*, with observations on the ocelli of some insects. Journ Morph. I, 1, pp. 193-226. Plate I.

———— 1889. Studies on the eyes of arthropods. II. Eyes of *Acilius* Jour. Morph. II, 1, pp. 97-190. Pl. VII-XIII, text figs. 1-5.

Petrunkewitsch, Alex. 1901. Die Richtungskörper and ihr Schicksal im befruchteten und unbefruchteten Bienenei. Zool. Jahrb. Abt. Anat. u. Ontog. Thiere. XIV, 4, pp. 573-608, Tafel 1-4.

———— 1902. Das Schicksal der Richtungskörper im Drohnenei. Ein Beitrag zur Kenntniss der natürlichen Parthenogenese. Zool. Jahrb. Abt. Anat. u. Ontog Thiere. XVII, 3, pp. 481-516 Tafel 1-3.

Pratt, H. S. 1900. The embryonic history of imaginal discs in *Melophagus ovinus* L., together with an account of the earlier stages in the development of the insect. Proc. Boston Soc. Nat. History. XXIX, 13, pp. 241-272. With 7 plates.

RABITO, LEONARDO. 1898. Sull' origine dell' intestino medio nella *Mantis religiosa.* Nat. Siciliano. II, 5-8, pp. 181-183.

RILEY, WILLIAM A. 1904. The embryological development of the skeleton of the head of *Blatta.* Thesis presented to the Faculty of Cornell University for the degree of Doctor of Philosophy. Amer. Nat. XXXVIII 455-456, pp. 777-810. 12 text figs. Printed as separate by Ginn and Co. Pub. The Athenaeum Press, Boston.

RITTER, RICHARD. 1890. Die Entwicklung der Geschlechtsorgane und des Darmes bei *Chironomus.* Ztschr. Wiss. Zool. L, 3, pp. 408-427. Mit Tafel XVI.

RIZZI, MARCO. 1912. Sullo sviluppo dell 'uovo di "*Bombyx (Sericaria) mori* L." nel primo mese deposizione. "Redia." VIII, 2, pp. 323-359. Con tavole II-V.

RÜCKERT, J. 1895. Ueber das Selbstständigbleiben der väterlichen und mütterlichen Kernsubstanz während der ersten Entwicklung des befruchteten *Cyclops* Eies. Arch. Mikros. Ant. XLV, 3.

SALING, THEODOR. 1907. Zur Kenntnis der Entwicklung der Keimdrüsen von *Tenebrio molitor* L. Mit 2 Tafeln und 14 Figuren im Text. Ztschr. Wiss. Zool. LXXXVI, 2, pp. 238-303.

SCHWANGART, F. 1904. Studien zur Entodermfrage bei den Lepidopteren. Ztschr. Wiss. Zool. LXXVI, 2, pp. 167-212. Mit Taf. XII und XIII und 4 Figuren im Text.

SCHWARTZE, DR. ERICH. 1899. Zur Kenntnis der Darmentwicklung bei Lepidopteren. Aus dem zoologischen Institute der Universität Berlin. Ztschr. Wiss. Zool. LXVI, 3, pp. 450-496. Taf. XXXI-XXXIV.

TANNREUTHER, GEO. W. 1907. History of the germ cells and early embryology of certain aphids. Zool. Jahrb. Abt. Anat., XXIV, 4, pp. 609-642. With plates 49-53.

TANQUARY, MAURICE COLE. 1913. Biological and embryological studies on Formicidae. III. Studies on the embryology of *Camponotus herculeanus* var *ferrugineus* Fabr. and *Myrmica scabrinodis* var *sabuleti* Meinert. Bull. Ill. State Lab. Nat. Hist., IX, art IX, pp. 454-477. With pl. LVII-LXIV.

TICHOMIROFF, A. 1879. Ueber die Entwicklungsgeschichte des Seidenwurms. Zool. Anz. II, 20, pp. 64-67.

———— 1882. [On the ontogeny of the *Bomby mori* within the egg.] (Russian), Izvyest. Imp. Obshch. Lyubit. Estest. Antrop. i. Ethnog. Moscow, XXXII, 4.

———— 1885. Chemische Studien über die Entwicklung der Insecteneier. Ztschr. Physiol. Chem. IX, 4-5, pp. 518-532.

———— 1890. Ueber die Entwicklung der *Calandra granaria.* Biol. Centbl. X, 13-14, p. 424.

TICHOMIROW, DR. ALEXANDER, 1892. Aus der Entwicklungsgeschichte der Insekten. Festeschrift zum 70. Geburtstage Leuckarts. Leipzig. pp. 337-346. Mit Tafel XXXV.

TICHOMIROWA, O. S. 1890. Zur Embryologie von *Chrysopa..* Biol. Centbl. X, 13-14, p. 423. (Der VIII Kongress russischer Naturforscher und Aertzte in St. Petersburg. II. Sektion für Zoologie. 1. Sitzung am 29. Dez. 1889. [9 Jan. 1890.]).

TIKHOMIROWA, OLGA. 1892. Sur l'histoire du dévelopment de *Chrysopoda perla* (L'Origine du mésoderme des cellules vitellines) Cong. Internat. Zool. 2, sess. 1. pp. 112-119.

TOYAMA, K. 1902. Contributions to the study of silk worms. I. On the embryology of the silk-worm. With plates VII-XI. Bul. Col. Agr. Tokyo Imp. Univ. V, pp. 73-118.

Tschuproff, Helene von. 1903. Ueber die Entwicklung der Keimblätter bei den Libellen. (Aus dem zoologischen Institut der Universität Berlin), (Volfräufige Mitteilung.) (Mit 1 Figur.) Zool. Anz. XXVII, pp. 29-34.

Uzel, Dr. Heinrich. 1897. Vörläufige Mittheilung über die Entwicklung der Thysanuren. Zool. Anz. XX, 528-529. pp. 125-132.

———— 1898. Studien über die Entwicklung der apterygoten Insecten. Mit 6 lith. Tafeln und 5 Abbildungen im Text. 58 pp.

Viallanes, H. 1887. Etudes histologiques et organologiques sur les centres nerveaux et les organes des sens des animaux articulés. 4te mémoire. Ann. Sci. Nat. Zool. II, pp. 5-100.

———— 1891. Sur quelques points de l'histoire du développement embryonnaire de la mante religieuse (*Mantis religiosa*). Ann. Sci. Nat. Zool. (7), XI, pp. 283-328. Con planches 12 et 13.

Voeltzkow, Alfred. 1888. Vorläufige Mittheilung über die Entwicklung im Ei von *Musca vomitoria*. Zool. Anz. XI, 278, pp. 235-236.

———— 1889. Entwickelung im Ei von *Musca vomitoria*. Arb. Inst. Wurzb. IX, pp. 1-48. Mit Tafeln I-IV.

———— 1889a. *Melolontha vulgaris*. Ein Beitrag zur Entwicklung im Ei bei Insecten. Arb. Inst. Wurzb. IX, pp. 49-64. Mit Tafel V.

Weismann, Dr. August. 1863. Die Entwicklung der Dipteren im Ei, nach Beobachtungen an *Chironomus* spec., *Musca vomitoria* und *Pulex canis*. Ztschr. Wiss. Zool. XIII, 1, pp. 107-158, u. 2, pp. 159-220. Mit Tafel VIII-XIII.

———— 1882. Beiträge zur Kenntniss der esten Entwicklungsvorgänge im Insektenei. Beiträge zur anatomie und embryologie als Festgabe Jacob Henle zum 4. April 1882. Bonn 1882. Hierzu Tafel X, XI, XII.

Wheeler, William Morton. 1889. The embryology of *Blatta germanica* and *Doryphora decemlineata*. Jour. Morph. III, 2, pp. 291-384. Plates XV-XX. 16 text figures.

———— 1891. Neuroblasts in the arthropod embryo. Jour. Morph. IV, 3, pp. 337-343.

———— 1892. Concerning the blood tissue of the insects. Parts I-III, 1 plate in No. 193. Psyche, VI, 190, 191, 193. Pp. 216-220, 233-236, 253-258.

———— 1893. A contribution to insect embryology. Jour. Morph. VIII, 1, pp. 1-160, pls. I-VI.

———— 1893a. The primitive number of Malpighian vessels in insects. Psyche, VI, pp. 457, 485, 497, 509, 539, 545, 561.

———— 1910. Ants, their structure, development and behavior. Columbia Univ. Press, N. Y. 1910.

Wielowiejski, Dr. Heinrich Ritter von. 1886. Ueber das Blutgewebe der Insekten. Eine vorlaufige Mittheilung. Ztschr. Wiss. Zool. XXXIV, 3, pp. 512-536.

Will, Ludwig. 1888. Zur Entwicklungsgeschichte der viviparen Aphiden. Biol. Centbl. VIII, 5, pp. 145-154.

———— 1888a. Entwicklungsgeschichte der viviparen Aphiden. Zool. Jahrb. Abth. Anat. Ontog. Thiere, III, pp. 201-286. Hierzu Tafel VI-IX.

Witlaczil, Dr. Emanuel. 1884. Entwicklungsgeschichte der Aphiden. Ztschr. Wiss. Zool. XL, 4, pp. 559-606. Mit Tafel XXVIII-XXXIV.

Woodworth, Charles W. 1889. Studies on the embryological development of *Euvanessa antiopa*. In: Scudder, Samuel Hubbard. Butterflies of New England, I, pp. 95-104, pl. 63.

INDEX

Names of authors are in small capitals, generic names in italics.

Figs. I-XV. Selected stages in the development of the egg. In the text the stage represented by each of the figures is designated by the Roman numeral of the corresponding figure. The drawings are made from eggs which have been fixed, stained, and mounted in balsam, and the eggs are therefore treated as semitransparent objects. The stages during the formation of the blastoderm are omitted, since nothing of importance is visible on the exterior of the egg, x.

Fig. I. Lateral view. The cleavage cells form a small group at the anterior end of the egg. The polar protoplasm (*PP*) is evident as a deep staining disk.

Fig. II. Lateral view. The cleavage cells have increased in number and form a conical figure.

Fig. III. Lateral view. The cleavage cells have further increased in number, and the conical figure formed by them has increased in size, so that in the anterior half of the egg the cleavage cells lie on the surface of the egg and are here beginning to form blastoderm.

III

II

PP

I

FIG. IV. Ventral view, 32-34 hours. The lateral folds (*LF*) have just made their appearance.

FIG V. Ventral view. About 36 hours. The lateral folds have greatly lengthened, and the lateral plates (*LP*) are approximating each other. The head fold of the amnion (*Am*) and the anterior mesenteron rudiment (*AMR*) are now visible.

FIG. VI. Ventral view. 38-42 hours. The lateral plates (*LP*) have become united along the greater portion of their length. The posterior mesenteron rudiment (*PMR*) is now visible.

IV

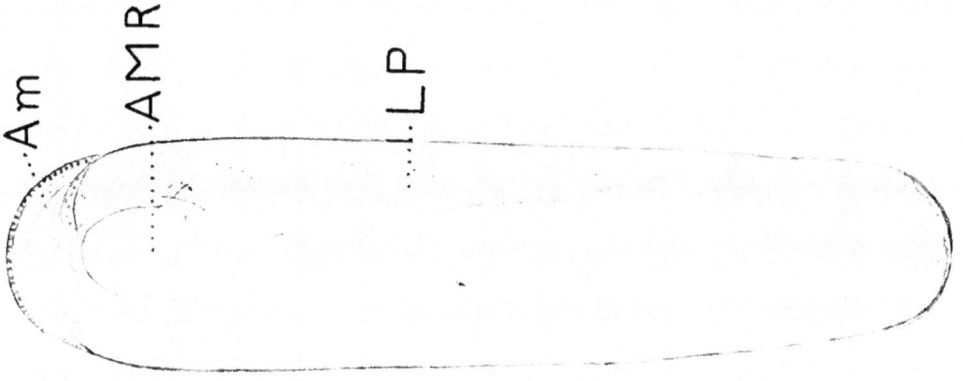

V

VI

Fig. VII. Ventral view. 42-44 hours. The lateral plates (*LP*) are almost completely united. The head fold of the amnion (*Am*) covers about one quarter of the ventral surface.

Fig. VIIa. Lateral view of the same egg as that shown in Fig. VII, showing the outlines of the germ band, and the procephalic lobes (*ProL*).'

Fig. VIII. Ventral view. 44-46 hours. Appearance of the appendages, silk glands, and tracheal system.

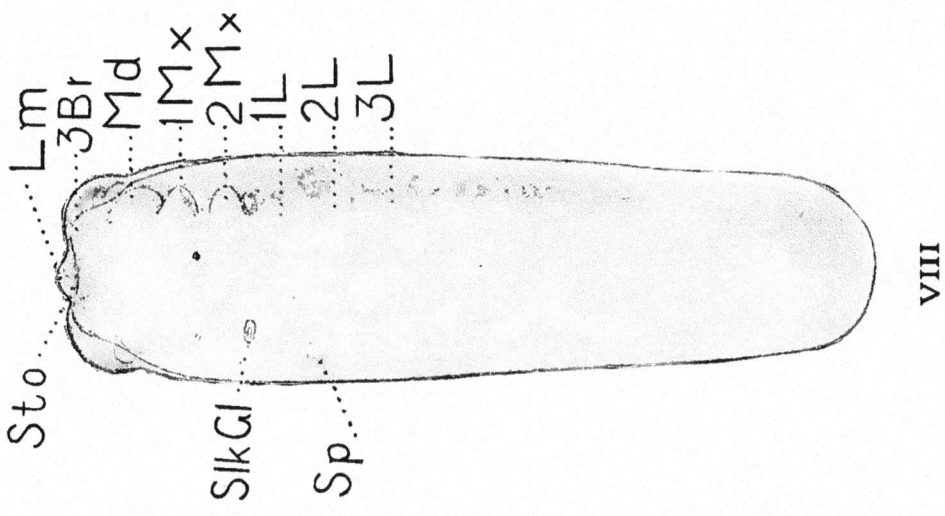

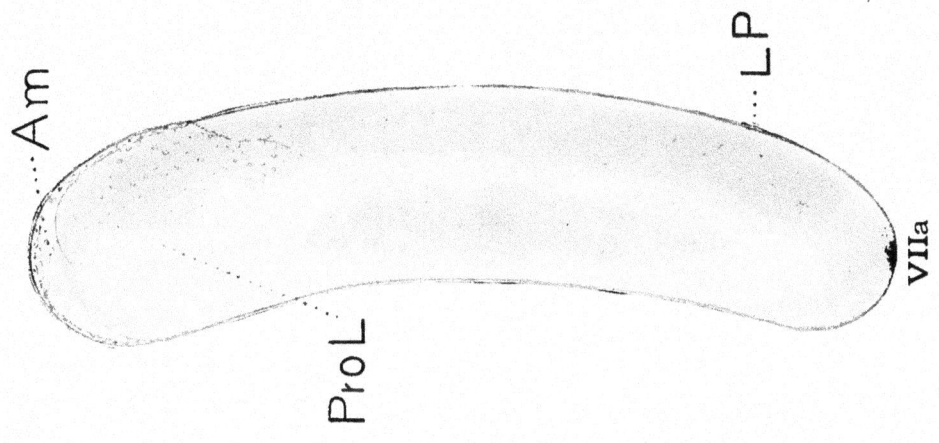

Lm
3Br
Md
1Mx
2Mx
1L
2L
3L

Sto

SlkGl

Sp

VIII

Am

ProL

LP

VIIa

Am

LP

VII

Fig. VIIIa.   Lateral view of the egg represented in Fig. VIII.

Fig. IX.   Ventral view.   48-50 hours.   The rudiments of the silk glands (*SlkGl*) are beginning to elongate.

Fig. X.   Ventral view, 52-54 hours.   The rudiments of the silk glands (*SklGl*) are about one-half of the length of the embryo.

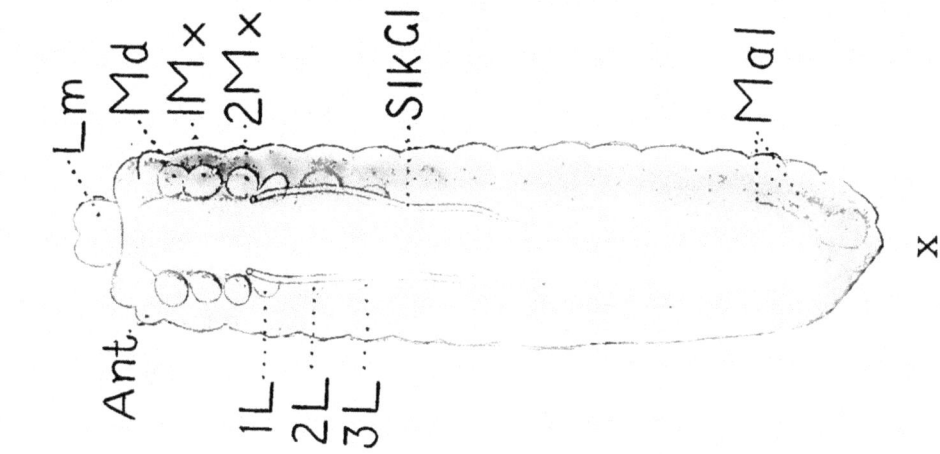

Lm
3Br
Ant
Md
1Mx
2Mx
SlkGl

Sp

1L
2L
3L

X

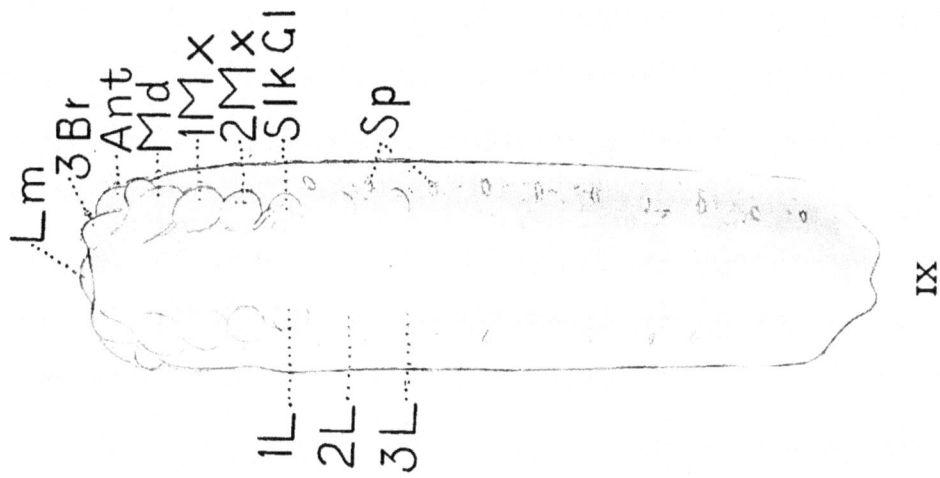

Lm
3Br
Ant
Md
1Mx
2Mx
SlkGl

Sp

1L
2L
3L

IX

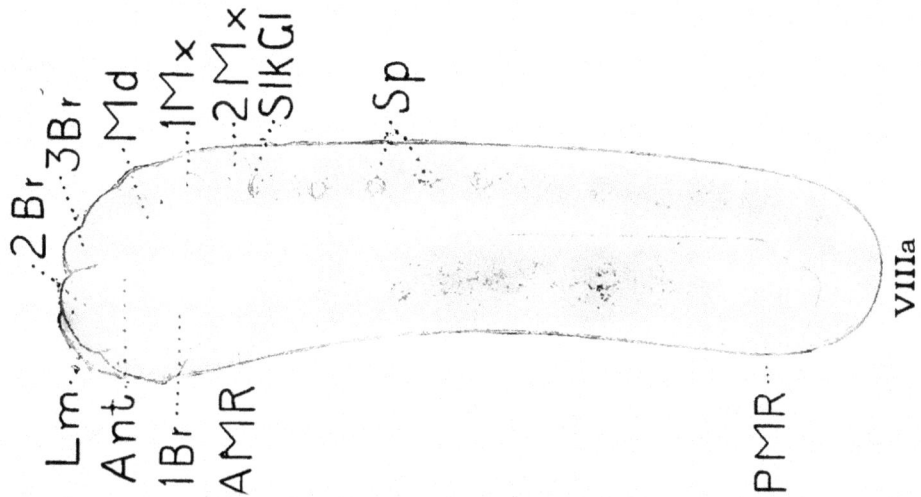

2Br
3Br
Lm
Ant
1Br
Md
1Mx
2Mx
SlkGl

Sp

AMR

PMR

VIIIa

FIG. Xa. Lateral view of the egg represented in Fig. X. The spiracles (*Sp*) have become contracted to minute circular apertures; the stomodaeum (*Sto*), proctodaeum (*Pro*), and the Malpighian tubules (*Mal*) are now evident.

FIG. XI. Ventral view. The silk glands (*SlkGl*) now reach into the posterior segments of the abdomen, the rudiments of the tracheal commissures (*TraCom*) of each are about to unite on the ventral mid-line.

FIG. XII. Ventral view. Estimated at 58-60 hours. The tracheal commissures (*TraCom*) are completed, the second maxillae (*2Mx*) are beginning to approach one another, and the ventral nerve cord (*VNC*) is visible from the surface.

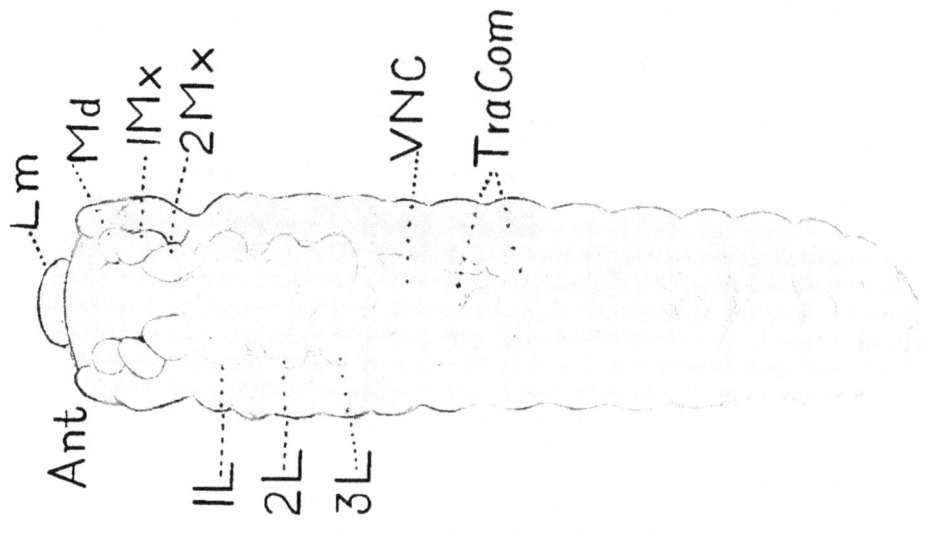

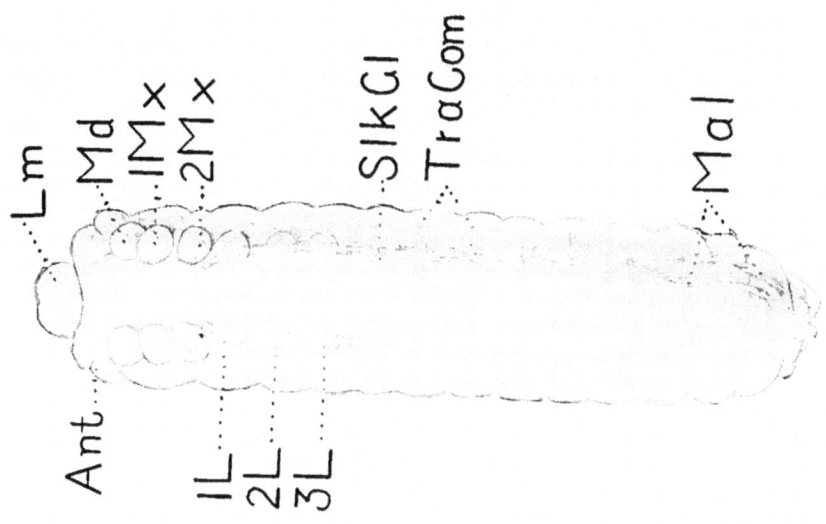

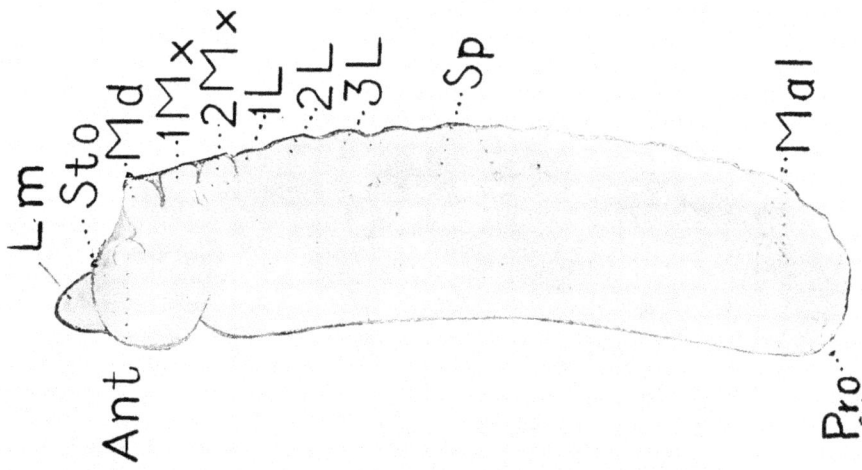

Fig. XIII. Ventral view. 66-68 hours. The second maxillae (*2Mx*) have begun to unite to form the labrum, bringing together the previously separate external openings of the silk glands (*SlkGl*).

Fig. XIV. Ventral view. 72-76 hours. Development virtually completed.

Fig. XV. Lateral view of newly hatched larva.

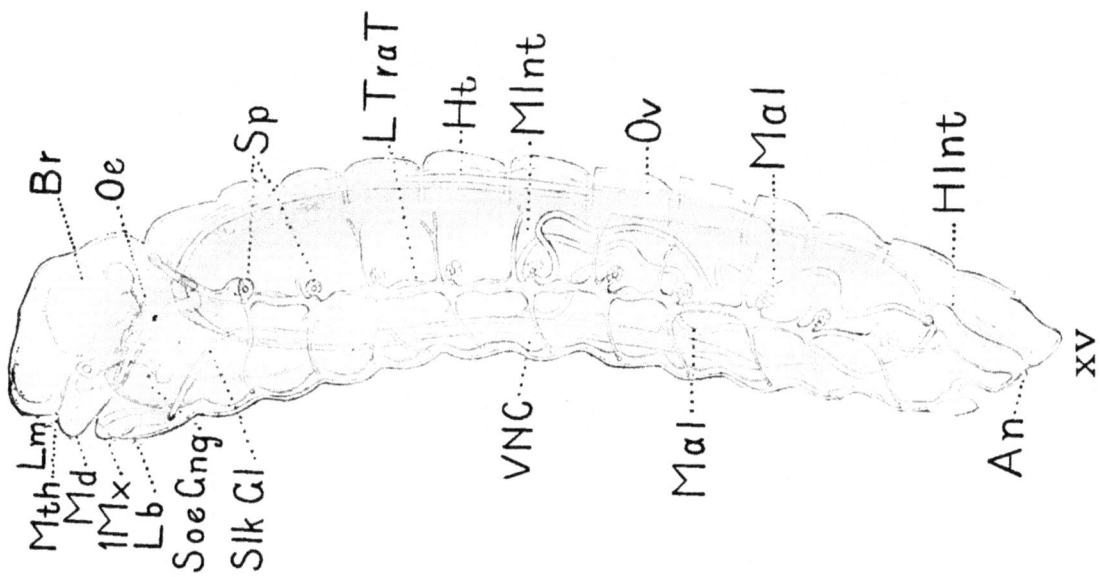

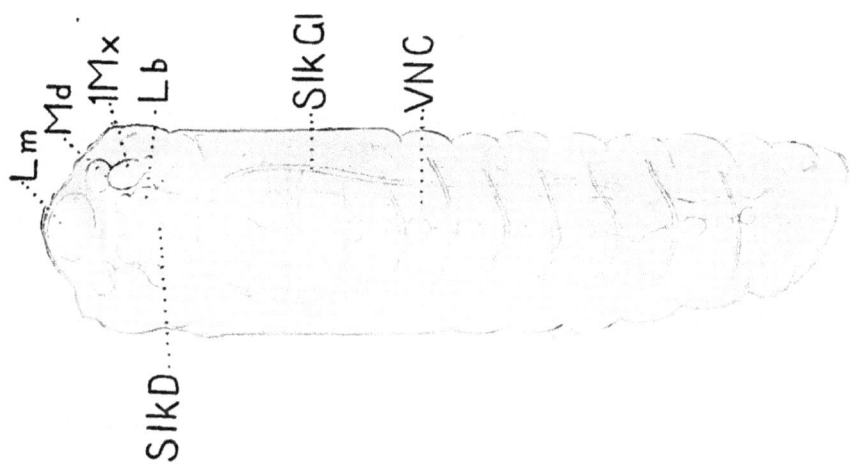

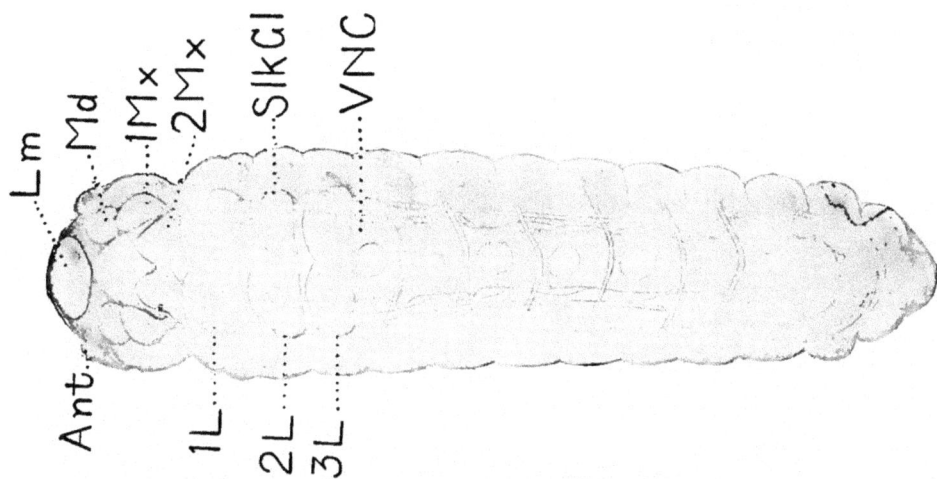

Lm

Br

Oe

Sp

LTraT

Ht

MInt

Ov

Mal

HInt

Mth

Md

1Mx

Lb

SoeCng

SlkGl

VNC

Mal

An

xv

Lm

Md

1Mx

Lb

SlkGl

VNC

SlkD

xiv

Lm

Md

1Mx

2Mx

SlkGl

VNC

Ant

1L

2L

3L

xiii

www.ingramcontent.com/pod-product-compliance
Lightning Source LLC
Chambersburg PA
CBHW051115200326
41518CB00016B/2514